Academic Preparation In Science

Second Edition

Teaching for Transition From High School To College

College Entrance Examination Board, New York, 1990

Academic Preparation in Science is one of a series of six books. The Academic Preparation Series includes books in English, the Arts, Mathematics, Science, Social Studies, and Foreign Language. Single copies of any one of these books can be purchased for $6.95. Orders for 5 through 49 copies of a single title receive a 20 percent discount; orders for 50 or more copies receive a 50 percent discount.

A boxed set of all the books in the Academic Preparation Series is available for $20. Orders for five or more sets receive a 20 percent discount. Each set also includes a copy of *Academic Preparation for College: What Students Need to Know and Be Able to Do.*

Payment or purchase order for individual titles or the set should be addressed to: College Board Publications, Box 886, New York, New York 10101-0886.

Figure 6. Reprinted by permission of Bolt, Beranek and Newman, Inc. from B. White and P. Horwitz. 1987. "Thinkertools: Enabling Children to Understand Physical Laws." Technical Report #6470. Figure 3.

Figure 8. Reprinted by permission of Lawrence Erlbaum Associates, Inc. from A. Whimbey and J. Lockhead. 1984. *Beyond Problem Solving and Comprehension.* p. 204.

Figure 9. Reprinted by permission from T.E. Raphael. 1987. "Research on Reading: But What Can I Teach on Monday?" Figure 2.1. In V. Richardson-Koehler, ed. *Educators' Handbook: A Research Perspective.*

ISBN: 0-87447-352-7

9 8 7 6 5 4 3 2

Contents

To Our Fellow Teachers of Science *1*

I. **Beyond the Green Book** *4*
Students at Risk, Nation at Risk *5*
The Classroom: At the Beginning as Well as the End of Improvement *7*
From Deficit to Development *9*
Dimensions for a Continuing Dialogue *10*

II. **Preparation and Outcomes** *14*
The Student as Learner *15*
The Teacher as Facilitator/Mediator *23*
Teaching the Process of Science: Laboratory and Mathematical Skills *27*
Learning Outcomes and the Structured Inquiry Approach... *31*
Approaching Scientific Questions Experimentally *36*
Gathering Scientific Information *38*
Organizing and Communicating Results *40*
Drawing Conclusions *42*
Recognizing the Role of Experimental Work in Constructing Theories *46*

III. **The Science Curriculum: Goals and Commentary** *51*
Integrating Content and Process in Science Instruction *52*
Higher Order Skills and Teaching Practice *55*
Summary Comments *58*
Beginning Science Education Earlier *59*
Encouraging Independent Learning *60*
Encourage the Use of All Available Resources *61*
Teach Skills Needed to Acquire Scientific Knowledge *62*
Build a Self-Consistent Knowledge Base *62*

IV. **Teaching Science** *63*
Kinematics *63*
Activity 1: Measuring the Acceleration Due to Gravity.... *64*

Activity 2: Probing and Dealing with a Common Misconception *68*
Respiration *70*
Chemical Properties and the Periodic Table *75*
Reasoning Qualitatively from a Historical Perspective *77*
What Kinds of Historical Events Might Students Use in Answering the Guiding Questions? *78*
Topics from Other Sciences *82*

V. **Science and the Basic Academic Competencies** *85*
Communication and Representation *85*
Speaking and Listening *86*
Writing and Representation *87*
Reading *92*
Observing *94*
Mathematics *98*
Reasoning *98*
Studying *100*
Using Computers *101*

VI. **Toward Further Discussion** *104*
Attracting, Preparing, and Retaining a Competent Cadre of Science Teachers *104*
Obstacles to Science Learning at the Elementary School Level *106*
Redressing Imbalances *108*
Effectiveness of Textbooks in Science Instruction *109*
Assessment of Learning Outcomes *109*
The Need for a Continuing Dialogue *111*

Bibliography *113*

Appendix *121*
A Biology Example of Structured Inquiry *121*
Approaching Scientific Questions Experimentally *121*
Gathering Scientific Information *123*
Organizing and Communicating Results *124*
Drawing Conclusions *125*
Recognizing the Role of Experimental Work in Constructing Theories *127*

Members of the Council on Academic Affairs, 1988–89 .... *130*

Figures

Figure 1 Heating curve for water (H_2O) 33
Figure 2 Cooling curve for paradichlorobenzene ($C_6H_4Cl_2$) 34
Figure 3 Graphical procedure for obtaining a "best fit," upper and lower limits to the gravitational acceleration data 67
Figure 4 Warburg respirometer 71
Figure 5 Burning candle inside a jar surrounded by water 73
Figure 6 Thinkertools display 90
Figure 7 Qualitative speed versus position graph from "terrain view" of road for a bicycle trip 92
Figure 8 Exercise to help students understand graphical representations 93
Figure 9 Examples of vocabulary mapping 95
Figure 10 Two common incorrect speed versus time graphs drawn by students depicting motion of a cart catapulted by a rubber band 97

Principal Writers

Jose P. Mestre, Associate Professor of Physics, University of Massachusetts, Amherst

Jack Lochhead, Director, Scientific Reasoning Research Institute, University of Massachusetts, Amherst

Science Advisory Committee, 1988—1989

Naomi Martin, Coordinator of Secondary Mathematics/Science, Horry County School District, Conway, South Carolina (*Chair*)

Victor M. De Leon, Dean of Academic Affairs, Hostos Community College, New York

Carole R. Goshorn, Chemistry Teacher, Columbus East High School, Columbus, Indiana

Jose P. Mestre, Associate Professor of Physics, University of Massachusetts, Amherst

Lois A. Peterson, Biology Teacher, El Cerrito High School, El Cerrito, California

Ronald L. Sass, Professor of Biology and Chemistry, Rice University, Houston, Texas

Arnold A. Strassenberg, Director, Teacher Preparation and Enhancement, National Science Foundation, Washington, D.C.

Acknowledgments

The College Board wishes to thank all the individuals and organizations that contributed to the preparation of this new edition of *Academic Preparation in Science*. In addition to those who served on the Science Advisory Committee and the Council on Academic Affairs, explicit acknowledgment should be accorded to a number of people. Fred Diehl and Jay Sylvester read early drafts of the manuscript and made many helpful suggestions. William Leonard also read the book in manuscript and helped us with corrections. Judith Goodenough made a major contribution to the preparation of the appendix. Finally, Adrienne Y. Bailey, vice president for Academic Affairs at the College Board, provided welcome support and encouragement throughout the project. Although none of these people is individually responsible for the content, this book is better and, we think, more useful because of their efforts.

Robert Orrill, Executive Editor

The College Board is a nonprofit membership organization committed to maintaining academic standards and broadening access to higher education. Its more than 2,600 members include colleges and universities, secondary schools, university and school systems, and education associations and agencies. Representatives of the members elect the Board of Trustees and serve on committees and councils that advise the College Board on the guidance and placement, testing and assessment, and financial aid services it provides to students and educational institutions.

The Educational EQuality Project is a 10-year effort of the College Board to strengthen the academic quality of secondary education and to ensure equality of opportunity for postsecondary education for all students. Begun in 1980, the project is under the direction of the Board's Office of Academic Affairs.

For more information about the Educational EQuality Project and inquiries about this report, write to the Office of Academic Affairs, The College Board, 45 Columbus Avenue, New York, New York 10023-6917.

To Our Fellow Teachers of Science

This second edition of *Academic Preparation in Science* is much changed from the previous edition published under the same title in 1986. With the exception of Chapter I, which is common to all six books of the Academic Preparation series, this edition has been almost entirely rewritten to reflect recent findings in cognitive research that brings us within reach of what Lauren Resnick has called a "new conception" of science education (Resnick 1983). These findings offer the possibility of developing new teaching strategies that will make science accessible to many more students. Indeed, as we will discuss in the pages that follow, a number of such strategies are already in place and are proving effective in helping students learn scientific concepts and in developing their problem-solving skills.

Few in the science community now doubt the need for major reform in science education. As stated in the report of Project 2061, *Science For All Americans*, "A cascade of recent studies has made it abundantly clear that both by national standards and international norms, U.S. education is failing to adequately educate too many students — and hence failing the nation" (American Association for the Advancement of Science 1989). Two myths have helped to create this emergency. The first is the widespread belief that the ability to learn science is possessed by only a select few. This is simply not true. Although not every student is capable of becoming a professional scientist, we strongly believe that individual motivation and effort, along with sound instruction, will permit most students to become scientifically literate. The second myth is that studying science primarily involves memorizing facts found in textbooks and that performing experiments is simply an exercise in verifying known phenomena. This view is perpetuated by the "learn-the-facts" approach common in much of science education.

Unfortunately, the thrust of recent educational "reform" in the sciences has been to increase the number of science courses required for graduation. The validity of this approach is doubtful.

It is difficult to see how improvement will come if all we do is more of what is now not being done very well. Moreover, there are indications that this "solution" may actually exacerbate the problem; those who like and do well in science courses will thrive while those who have been struggling with their courses will drop out of science and out of school. Thus, we do not share the view that "more of the same" will improve the current state of science education. What we need is a "new conception" of science instruction.

This new conception should have three overarching goals, regardless of the student's grade level or ability.

1. To instill in all students the idea that science is a process of inquiry and problem solving.
2. To provide all students with a sufficient working knowledge of science to be able to deal effectively with the critical technological-societal issues that confront us today.
3. To strengthen preparation for the next generation of scientists who are crucial to maintaining our international position in science and our national well-being.

We propose a dual approach in tackling these ambitious goals. Not only should we help students to gain a solid grounding in the process of inquiry that is indispensable to scientific reasoning but we must incorporate into this instruction those recent findings from cognitive research that enable students to learn most effectively. In classroom practice, this dual approach results in three themes or emphases that readers will find recurring throughout this book: the process of "doing" science (i.e., the scientific, or experimental, paradigm); the student as learner and problem solver; and the teacher as facilitator. We are hopeful that science teachers will share and help in shaping this new conception of teaching and learning science.

The chapters that follow are a first step in achieving the goals outlined above. Chapter I describes the relationship of this book to the Educational Equality Project and its publication, *Academic Preparation for College* (College Board 1983).

Chapter II begins our discussion of the cognitive processes involved in learning science and relates these to the learning outcomes we hope all students will achieve. The chapter continues with an exploration of how teachers can facilitate students' learning science and concludes with an illustration of a classroom approach we call "structured inquiry."

The focus of Chapter III is the science curriculum. In this chapter, we discuss the importance of integrating content and process in science instruction, helping students to learn independently, and devising an effective agenda for beginning science instruction early in the students' schooling.

Chapter IV examines the experimental paradigm in greater detail and provides specific examples from physics, biology, and chemistry that further develop the structured inquiry approach to teaching and learning science.

Chapter V contains a discussion of the relationship between science instruction and the "Basic Academic Competencies," those broad intellectual skills that students need for success in all fields of study.

Chapter VI concludes this volume with a discussion of broad-based issues of significance to science teachers. These include training and maintaining a qualified cadre of science teachers; assessing the outcomes of learning; and maintaining an ongoing, stimulating dialogue among science teachers.

In addition, we provide a comprehensive bibliography and an appendix describing how the "structured inquiry" approach might be applied to instruction in biology.

Finally, we urge all science teachers to regard this book as a catalyst for further discussion rather than as a manual for prescriptive science instruction. The ideas will have little impact unless they are explored and refined by classroom teachers. If this new conception of science education is to have effect, it must begin with you, the practitioner.

I. Beyond the Green Book

Identifying the academic preparation needed for college is a first step toward providing that preparation for all students who might aspire to higher education. But the real work of actually achieving these learning outcomes lies ahead. (College Board 1983, 31).

This book is a sequel to *Academic Preparation for College: What Students Need to Know and Be Able to Do,* which was published in 1983 by the College Board's Educational EQuality Project. Now widely known as the Green Book, *Academic Preparation for College* outlined the knowledge and skills students need in order to have a fair chance at succeeding in college. It summarized the combined judgments of hundreds of educators in every part of the country. The Green Book sketched learning outcomes that could serve as goals for high school curriculums in six Basic Academic Subjects: English, the arts, mathematics, science, social studies, and foreign languages. It also identified six Basic Academic Competencies on which depend, and which are further developed by, work in these subjects. Those competencies are reading, writing, speaking and listening, mathematics, reasoning, and studying. The Green Book also called attention to additional competencies in using computers and observing, whose value to the college entrant increasingly is being appreciated.

With this book we take a step beyond *Academic Preparation for College.* The Green Book simply outlined desired results of high school education—the learning all students need to be adequately prepared for college. It contained no specific suggestions about how to achieve those results. Those of us working with the Educational EQuality Project strongly believed—and still believe—that ultimately curriculum and instruction are matters of local expertise and responsibility. Building consensus on goals, while leaving flexibile the means to achieve them, makes the most of educators' ability to respond appropriately and creatively to conditions in their own schools. Nevertheless, teachers and administrators, particularly those closely associated with the EQuality project, often have asked how the outcomes sketched in the Green Book might be translated into actual curriculums and instructional practices—how they can get on with the "real work" of education. These requests in part seek suggestions about how the Green Book goals might be achieved; perhaps to an even greater extent they express a desire to get a fuller picture of those very briefly stated goals. Educators prefer to think realistically, in terms of courses and lessons. Discussion of proposals such as those in the Green Book proceeds more readily when goals are filled out and cast into the practical language of possible courses of action.

To respond to these requests for greater detail, and to encourage further nationwide discussion about what should be happening in our high school classrooms, teachers working with the Educational EQuality Project have prepared this book and five like it, one in each of the Basic Academic Subjects. By providing suggestions about how the outcomes described in *Academic Preparation for College* might be achieved, we hope to add more color and texture to the sketches in that earlier publication. We do not mean these suggestions to be prescriptive or definitive, but to spark more detailed discussion and ongoing dialogue among our fellow teachers who have the front-line responsibility for ensuring that all students are prepared adequately for college. We also intend this book and its companions for guidance counselors, principals, superintendents, and other officials who must understand the work of high school teachers if they are better to support and cooperate with them.

Students at Risk, Nation at Risk

Academic Preparation for College was the result of an extensive grassroots effort involving hundreds of educators in every part of the country. However, it was not published in a vacuum. Since the beginning of this decade, many blue-ribbon commissions and studies also have focused attention on secondary education. The concerns of these reports have been twofold. One, the reports note a perceptible decline in the academic attainments of students who graduate from high school, as indicated by such means as standardized test scores and comments from employers; two, the reports reflect a widespread worry that, unless students are better educated, our national welfare will be in jeopardy. *A Nation at Risk* made this point quite bluntly:

> Our Nation is at risk. Our once unchallenged preeminence in commerce, industry, science, and technological innovation is being overtaken by competitors throughout the world. . . . The educational foundations of our society are presently being eroded by a rising tide of mediocrity that threatens our very future as a Nation and a people (National Commission on Excellence in Education 1983, 5).

The Educational EQuality Project, an effort of the College Board throughout the decade of the 1980s to improve both the quality of preparation for college and the equality of access to it, sees another aspect of risk: if our nation is at risk because of the level of students' educational attainment, then we must be more concerned with those students who have been most at risk.

Overall, the predominance of the young in our society is ending. In 1981, as the EQuality project was getting under way, about 41 percent of our country's population was under 25 years old and 26 percent was 50 years old or older. By the year 2000, however, the balance will have shifted to 34 percent and 28 percent, respectively. But these figures do not tell the whole story, especially for those of us working in the schools. Among certain groups, youth is a growing segment of the population. For example, in 1981, 71 percent of black and 75 percent of Hispanic households had children 18 years old or younger. In comparison, only 52 percent of all white households had children in that age category. At the beginning of the 1980s, children from minority groups already made up more than 25 percent of all public school students (Boyer 1983, 4–5; U.S. Department of Education 1983, 43). Clearly, concern for improving the educational attainments of all students increasingly must involve concern for students from such groups of historically disadvantaged Americans.

How well will such young people be educated? In a careful and thoughtful study of schools, John Goodlad found that "consistent with the findings of virtually every study that has considered the distribution of poor and minority students . . . minority students were found in disproportionately large percentages in the low track classes of the multiracial samples [of the schools studied]" (Goodlad 1984, 156). The teaching and learning that occur in many such courses can be disappointing in comparison to that occurring in other courses. Goodlad reported that in many such courses very little is expected, and very little is attempted (Goodlad 1984, 159).

When such students are at risk, the nation itself is at risk, not only economically but morally. That is why this book and its five companions offer suggestions that will be useful in achieving academic excellence for *all* students. We have attempted to take into account that the resources of some high schools may be limited and that some beginning high school students may not be well prepared. We have tried to identify ways to keep open the option of preparing adequately for college as late as possible in the high school years. These books are intended for work with the broad spectrum of high school students—not just a few students and not only those currently in the "academic track." We are firmly convinced that many more students can—and, in justice, should—profit from higher education and therefore from adequate academic preparation.

Moreover, many more students actually enroll in postsecondary education than currently follow the "academic track" in high school. Further, discussions with employers have emphasized that many of the same academic competencies needed by college-bound students also are needed by high school students going directly into the world of work. Consequently, the Educational EQuality Project, as its name indicates, hopes to contribute to achieving a democratic excellence in our high schools.

The Classroom: At the Beginning as Well as the End of Improvement

A small book such as this one, intended only to stimulate dialogue about what happens in the classroom, cannot address all the problems of secondary education. On the other hand, we believe that teachers and the actual work of education—that is to say, curriculum and instruction—should be a more prominent part of the nationwide discussion about improving secondary education.

A 1984 report by the Education Commission of the States found that 44 states either had raised high school graduation requirements or had such changes pending. Twenty-seven states had enacted new policies dealing with instructional time, such as new extracurricular policies and reduced class sizes (Task Force on Education for Economic Growth 1984, 27). This activity reflects the momentum for and concern about reform that has been generated recently. It demonstrates a widespread recognition that critiques of education without concrete proposals for change will not serve the cause of improvement. But what will such changes actually mean in the classroom? New course requirements do not necessarily deal with the academic quality of the courses used to fulfill those requirements. Certain other kinds of requirements can force instruction to focus on the rote acquisition of information in the exclusion of fuller intellectual development. Manifestly, juggling of requirements and courses without attention to what needs to occur between teachers and students inside the classroom will not automatically produce better prepared students. One proponent of reform, Ernest Boyer, has noted that there is danger in the prevalence of "quick-fix" responses to the call for improvement. "The depth of discussion about the curriculum . . . has not led to a serious and creative look at the nature of the curriculum. . . . states [have not asked] what we ought to be teaching" (Boyer 1984, 33).

Such questioning and discussion is overdue. Clearly, many improvements in secondary education require action outside the classroom and the school. Equally clearly, even this action should be geared to a richer, more developed understanding of what is needed in the classroom. By publishing these books we hope to add balance to the national debate about improving high school education. Our point is not only that it is what happens between teachers and students in the classroom that makes the difference. Our point is also that what teachers and students do in classrooms must be thoughtfully considered before many kinds of changes, even exterior changes, are made in the name of educational improvement.

From Deficit to Development

What we can do in the classroom in limited, of course, by other factors. Students must be there to benefit from what happens in class. Teachers know firsthand that far too many young people of high school age are no longer even enrolled. Nationally, the drop-out rate in 1980 among the high school population aged 14 to 34 was 13 percent. It was higher among low-income and minority students. Nearly 1 out of 10 high schools had a drop-out rate of over 20 percent (U.S. Department of Education 1982, 68; Rock 1984, 4).

Even when students stay in high school, we know that they do not always have access to the academic preparation they need. Many do not take enough of the right kinds of courses. In 1980, in almost half of all high schools, a majority of the students in each of those schools was enrolled in the "general" curriculum. Nationwide, only 38 percent of high school seniors had been in an academic program; another 36 percent had been in a general program; and 24 percent had followed a vocational/technical track. Only 39 percent of these seniors had enrolled for three or more years in history or social studies; only 33 percent had taken three or more years of mathematics; barely 22 percent had taken three or more years of science; and less than 8 percent of these students had studied Spanish, French, or German for three or more years (U.S. Department of Education 1982, 70).

Better than anyone else, teachers know that, even when students are in high school and are enrolled in the needed academic courses, they must attend class regularly. Yet some school systems report daily absence rates as high as 20 percent. When 1 out of 5 students enrolled in a course is not actually there, it is difficult even to begin carrying out a sustained, coherent program of academic preparation.

As teachers we know that such problems cannot be solved solely by our efforts in the classroom. In a world of disrupted family and community structures; economic hardship; and rising teenage pregnancy, alcoholism, and suicide, it would be foolish to believe that attention to curriculum and instruction can remedy all the problems leading to students' leaving high school, taking the wrong courses, and missing classes. Nonetheless, what happens in the high school classroom—once students are there—is important in preparing students not only for further education but for life.

Moreover, as teachers, we also hope that what happens in the classroom at least can help students stick with their academic work. Students may be increasingly receptive to this view. In 1980 more than 70 percent of high school seniors wanted greater academic emphasis in their schools; this was true of students in all curriculums. Mortimer Adler may have described a great opportunity:

> There is little joy in most of the learning they [students] are now compelled to do. Too much of it is make-believe, in which neither teacher nor pupil can take a lively interest. Without some joy in learning—a joy that arises from hard work well done and from the participation of one's mind in a common task—basic schooling cannot initiate the young into the life of learning, let alone give them the skill and the incentive to engage in it further (Adler, 1982, 32).

Genuine academic work can contribute to student motivation and persistence. Goodlad's study argues strongly that the widespread focus on the rote mechanics of a subject is a surefire way to distance students from it or to ensure that they do not comprehend all that they are capable of understanding. Students need to be encouraged to become inquiring, involved learners. It is worth trying to find more and better ways to engage them actively in the learning process, to build on their strengths and enthusiasms. Consequently, the approaches suggested in these books try to shift attention from chronicling what students do not know toward developing the full intellectual attainments of which they are capable and which they will need in college.

Dimensions for a Continuing Dialogue

This book and its five companions were prepared during 1984 and 1985 under the aegis of the College Board's Academic Advisory Committees. Although each committee focused on the particular issues facing its subject, the committees had common purposes and common approaches. Those purposes and approaches may help give shape to the discussion that this book and its companions hope to stimulate.

Each committee sought the assistance of distinguished writers and consultants. The committees considered suggestions made in the dialogues that preceded and contributed to *Academic Preparation for College* and called on guest colleagues for further suggestions and insights. Each committee tried to take account of the best available thinking and research but did not merely pass along the results of research or experience. Each deliberated about those findings and then tried to suggest approaches that had actually worked to achieve learning outcomes described in *Academic Preparation for College*. The suggestions in these books are based to a great extent on actual, successful high school programs.

These books focus not only on achieving the outcomes for a particular subject described in the Green Book but also on how study of that subject can develop the Basic Academic Competencies. The learning special to each subject has a central role to play in preparing students for successful work in college. That role ought not to be neglected in a rush to equip students with more general skills. It is learning in a subject that can engage a student's interest, activity, and commitment. Students do, after all, read about *something*, write about *something*, reason about *something*. We thought it important to suggest that developing the Basic Academic Competencies does not replace, but can result from, mastering the unique knowledge and skills of each Basic Academic Subject. Students, particularly hungry and undernourished ones, should not be asked to master the use of the fork, knife, and spoon without being served an appetizing, full, and nourishing meal.

In preparing the book for each subject, we also tried to keep in mind the connections among the Basic Academic Subjects. For example, the teaching of English and the other languages should build on students' natural linguistic appetite and development—and this lesson may apply to the teaching of other subjects as well. The teaching of history with emphasis on the full range of human experience, as approached through both social and global history, bears on the issue of broadening the "canon" of respected works in literature and the arts. The teaching of social studies, like the teaching of science, involves mathematics not only as a tool but as a mode of thought. There is much more to make explicit and to explore in such connections among the Basic Academic Subjects. Teachers may teach in separate departments, but students' thought is probably not divided in the same way.

Finally, the suggestions advanced here generally identify alternate ways of working toward the same outcomes. We wanted very much to avoid any hint that there is one and only one way to achieve the outcomes described in *Academic Preparation for College*. There are many good ways of achieving the desired results, each one good in its own way and in particular circumstances. By describing alternate approaches, we hope to encourage readers of this book to analyze and recombine alternatives and to create the most appropriate and effective approaches, given their own particular situations.

We think that this book and its five companion volumes can be useful to many people. Individual teachers may discover suggestions that will spur their own thought about what might be done in the classroom; small groups of teachers may find the book useful in reconsidering the science program in their high school. It also may provide a takeoff point for in-service sessions. Teachers in several subjects might use it and its companions to explore concerns, such as the Basic Academic Competencies, that range across the high school curriculum. Principals may find these volumes useful in refreshing the knowledge and understanding on which their own instructional leadership is based.

We also hope that these books will prove useful to committees of teachers and officials in local school districts and at the state level who are examining the high school curriculum established in their jurisdictions. Public officials whose decisions directly or indirectly affect the conditions under which teaching and learning occur may find in the books an instructive glimpse of the kinds of things that should be made possible in the classroom.

Colleges and universities may find in all six books occasion to consider not only how they are preparing future teachers but also whether their own curriculums will be suited to students receiving the kinds of preparation these books suggest. But our greatest hope is that this book and its companions will be used as reference points for dialogues between high school and college teachers. It was from such dialogues that *Academic Preparation for College* emerged. We believe that further discussions of this sort can provide a wellspring of insight and energy to move beyond the Green Book toward actually providing the knowledge and skills all students need to be successful in college.

We understand the limitations of the suggestions presented here. Concerning what happens in the classroom, many teachers, researchers, and professional associations can speak with far greater depth and detail than is possible in the pages that follow. This book aspires only to get that conversation going, particularly across the boundaries that usually divide those concerned about education, and especially as it concerns the students who often are least well served. Curriculum, teaching, and learning are far too central to be omitted from the discussion about improving education.

II. Preparation and Outcomes

Teaching students science is a challenging task. Unfortunately, there is overwhelming evidence that, at present, we are not as effective as we can and must be. Report after report confirms that most Americans are not scientifically literate and that U.S. students rank near the bottom in international studies of educational performance in science and mathematics (International Association for the Evaluation of Educational Achievement 1988). This is alarming because the well-being of the nation depends, to a great extent, on our competence in science.

Any attempt to improve on this situation must begin by considering how students learn science—and how they fail to do so. Until recently, little research existed on this topic. Today, findings from numerous research studies have challenged common assumptions about how science is learned (Driver 1983). For example, the assumption that students come to us as "blank slates" which we then fill with scientific knowledge is no longer tenable; we know that students' previous experiences play a critical role in what students learn, or fail to learn, in our classrooms (Resnick 1983). Moreover, we also have learned from this research that making a teaching presentation more lucid is in itself insufficient to help students when they fail to understand a lesson on some scientific concept (Mestre 1987). A review of these research findings and their implications for instruction will be discussed below.

We must also consider current emphases in the teaching of science and ask whether these need to be changed. Perhaps teaching science is unlike teaching other subjects in that the teacher must simultaneously address three related areas: scientific concepts, problem solving skills, and laboratory skills. In teaching scientific concepts and problem solving skills, we tend to overemphasize the formal or mathematical (quantitative) aspects of scientific reasoning at the expense of the equally important, but more informal, qualitative aspects. We often find students who are capable of completing complex calculations, such as calculating the path of a projectile, but who cannot identify the concept underlying the calculation, such as the force acting on the projectile (Clement 1982). Research into how scientists actually think and solve problems suggests that they rely extensively on qualitative reasoning both in planning strategies for solving problems and in resolving possible conflicts in their strategies. Techniques for encouraging qualitative reasoning will be discussed below and in Chapter IV.

Perhaps the one area in science instruction where a change of emphasis is most urgently needed is in teaching the experimental paradigm, also known as "the process of science." This process involves observing, forming hypotheses, designing experiments to test hypotheses, analyzing and distilling data, modeling, and theory building. Rather than teaching students the experimental paradigm as a process of inquiry, we often convey to students that "doing" science consists of performing routine laboratory work to verify known phenomena. Here lies our greatest challenge as teachers of science. We must not only convey to students that scientific knowledge is acquired through a process of inquiry but also instill in them confidence that this process will allow them to successfully answer carefully posed questions. To accomplish this, we must help students realize that knowing the answer to a particular scientific question is not nearly as important as devising a procedure that will lead us to the answer.

These issues—the student as learner, the teacher as mediator between a body of scientific knowledge and the student, and the teaching of the experimental paradigm—will be revisited in the next three sections in increasing detail. We conclude this chapter with an examination of the crucial elements involved in teaching the process of scientific inquiry.

The Student as Learner

Recent cognitive research indicates that students enter our science classrooms possessing "naive theories" that they use to explain real-world events (Resnick 1983; Mestre 1987; Carey 1986; Linn 1986; Helm and Novak 1983; Novak 1987). Students actively, albeit unconsciously, construct such theories in an attempt to organize and make sense out of the world around them. Although attempts to categorize and explain phenomena by constructing theories is at the heart of scientific inquiry, there are several problems with the naive theories that students construct. These theories are often incomplete, fragmented, and fraught with misconceptions. As a result, they can interfere with learning.

Unfortunately, many of these misconceptions are not easily changed. Because students have spent considerable time and energy constructing their naive theories, they have an emotional and intellectual investment in them. Research consistently shows that students embrace their erroneous beliefs tenaciously and often explain away what directly conflicts with their naive theories by either reinterpreting an event or by making inconsequential modifications to their theories. The most convincing evidence that misconceptions are deep-rooted and difficult to dislodge is that many persist even after the student completes a course of instruction taught by a very competent teacher (Halloun and Hestenes 1987). These observations discredit the assumption that learning science depends almost solely on the clarity of the presentation.

Some examples from various science disciplines will show how pervasive naive theories are. In biology, for example, studies of students' understanding of genetics (Kinnear 1983) and photosynthesis (Wandersee 1983) have revealed several misconceptions and tendencies to apply rote rules inappropriately. In an example of the latter, freshman college students were given a genetic breeding situation in which a green bird (green color dominant) was crossed with a blue bird (blue color recessive) resulting in eight fledglings in the ratio of three green birds for each blue. Then they were asked to judge the accuracy of the statement: "This result proves that the two parent birds were heterozygous." Although these students had shown an understanding of the concept of heterozygosity and homozygosity, over one-half of the freshmen incorrectly labeled this statement "true." All of these students had had a high school biology course with a genetics component. Yet the students' attempt to solve this genetics problem simply elicited a rote response when they were prompted by the ratio 3:1. Indeed, many of the students who gave the wrong response immediately realized their error when they were asked to rework the problem by starting with writing down the parental genotypes.

A study of students' understanding of how plants made food, which included large samples of students from grades 5, 8, and 11, as well as college sophomores, revealed many misconceptions (Wandersee 1983). This study used a test with both multiple-choice and free-response items to investigate students' understanding of the role of soil in plant growth, the role of photosynthesis in the carbon cycle, the roles of the leaf and light energy in photosynthesis, and the primary source of food in green plants. Although students at higher grade levels gave a higher percentage of correct responses, a significant number of students at all grade levels displayed a wide range of misconceptions: the soil loses weight as plants grow in it, the soil is the plant's food, roots absorb soil, plants convert energy from the sun directly into matter, plants give off mainly carbon dioxide, the leaf's main job is to capture rain and water vapor in the air, plants get their food from the roots and store it in their leaves, chlorophyll is the plant's blood, and chlorophyll is no longer available in the air during autumn and winter so the leaf cannot then get food.

The richest body of literature on students' naive or alternative conceptions is in physics (McDermott 1984; McCloskey, Caramazza, and Green 1980; Champagne, Klopfer, and Gunstone 1982; Clement 1982; Fredette and Clement 1981; Goldberg and McDermott 1986; Minstrell 1982). The world in which we live is, in a sense, a large physics laboratory in which experiments (physical events) are continuously being performed for our observation. In observing events, we construct naive theories during the course of our lives to explain physical phenomena. Let us look at two examples from mechanics.

In the first example, students were asked to indicate each force acting on a coin that is tossed straight up in the air and has reached the halfway point in its trajectory. Students were to indicate each force with an arrow, the length of the arrow denoting the magnitude, and the arrowhead pointing in the direction of the force. Nearly 90 percent of a group of college freshmen majoring in engineering, who had not yet taken an introductory college-level mechanics course, answered incorrectly. Most students answering incorrectly indicated two forces, one pointing downward representing the force of gravity, another pointing upward representing "the original upward force of the hand." The appropriate answer is that only the gravitational force is present while the coin is airborne (neglecting the insignificant effect of air resistance). What is incomprehensible to many people is that an object can continue to move in one direction when the only net force acting on the object is in the opposite direction. What is surprising to most instructors is that about 70 percent of a sample of engineering majors who had *finished* their college-level mechanics course, when asked the same question, had the same misconception (Clement 1982).

The second example involves curvilinear motion. When college students, many of whom had taken physics, were asked to draw the path taken by a ball that had emerged from a hollow spiral tube, a large percentage drew a curved path that followed the general shape of the spiral. Those who made this error claimed that the ball had the ability to acquire a "force" or "momentum" that caused it to continue moving in a curved path after emerging from the tube (McCloskey 1983; McCloskey, Caramazza, and Green 1980). Yet this conflicts with observation. Once it is outside the tube, the ball will move in a straight line because there is no longer any force restricting the motion to a curved path.

A fairly recent body of research has uncovered several interesting misconceptions in astronomy and cosmology. In the first large national survey of astronomical knowledge, 1,120 adults were asked to answer several multiple-choice questions (Lightman, Miller, and Leadbeater 1987). The first question asked whether the sun is a planet, a star, or something else. Only 55 percent of those surveyed correctly stated that the sun is a star. In response to the question, "Is the universe getting bigger or getting smaller, or does it remain the same size?" only 24 percent responded that the universe is expanding—common knowledge in cosmology for more than 60 years. Of the 200 adults who stated they would be troubled if they found out that the universe was indeed expanding, 184 (or 92 percent) were concerned that the expansion might present a danger to earth. That younger and better-educated adults were the most knowledgeable may attest to an improvement in instruction in astronomy. However, negative emotional reactions to the discovery of an expanding universe, such as fear of unknown change, fear of loss of control, a sense of helplessness, feelings of insignificance, and concern over immediate danger to the earth and possible death, indicate the delicate relationship between deeply held beliefs and science education.

A second study of misconceptions about astronomy revealed that an alarming number of high school students believed that it is hot in the summer but cold in the winter because the sun is closer to the earth in the summer than in the winter (Sadler 1987). Answers to a second question asking why it is dark at night and light during the day included: the moon blocks out the sun, the sun goes out at night, and the atmosphere blocks out the sun at night. Over half the students who participated in the study were completing an earth science course in which 25 percent of the subject matter was astronomy, yet these students were unable to answer these two simple questions correctly, although they did use more technical jargon in their answers than those who had not taken the course.

One of the most revealing findings of this study was the shock expressed by an earth science teacher after seeing her students' responses. Because she had recently completed lessons on these topics, this teacher had predicted scores ranging from 7 to 10 (on a scale of 1 to 10) for a sample of her students on the day-night question and on a question about the phases of the moon. When shown videotaped interviews of these students answering the two questions, she dropped her assigned scores an average of six points; she also displayed considerable dismay over the "wild" views still held by some of her best students, despite classroom instruction.

The existence of these misconceptions is perhaps less frightening than that they have been ignored for so long. A view of science that stresses only formal theory seldom offers the opportunity to consider the qualitative contexts in which such misconceptions have an opportunity to surface. Some scientists have dismissed misconceptions as inconsequential to learning science. We believe their view fails to recognize the role that qualitative knowledge plays in sophisticated thinking. Studies of how experts solve problems show that qualitative, conceptual understanding is critical for organizing an intelligent, rational attack on any problem. The highly abstract nature of the qualitative understanding underlying much of modern physics has misled many physicists and mathematicians into believing that it is possible to function on a purely formal level. However, this myth is contradicted in the writings of prominent scientists such as Richard Feynman, Stephen Hawking, and Albert Einstein, who show clearly the role that qualitative thinking played in their work. Furthermore, studies of problem solving by students reveal that it is precisely because students lack such skills that they are unable to solve nonstandard problems and do not apply what they know effectively in novel situations.

In addition to the question, "What are the common naive theories that students construct that interfere with learning?" is the question, "How do students acquire expertise?" Most of our understanding comes from studies comparing the performance of "experts" with that of "novices" in tasks that probe the way scientific knowledge is stored in human memory and how this knowledge is used in solving problems. Close attention to findings from this body of research can help us gain insights to shape instruction that facilitate the acquisition of expertise.

Cognitive research reveals that all of us store knowledge in long-term memory in clusters (called "chunks") with the information in each cluster related by some underlying theme called "schema." Knowledge stored in isolation is quickly forgotten or becomes inaccessible (Resnick 1983). Although we all use these clusters to organize knowledge, experts and novices organize knowledge quite differently from one another. The expert's knowledge is structured in a hierarchical network of clusters. Categories within the network are often based on qualitative rather than quantitative distinctions. The higher a cluster is in the hierarchy, the more overarching the theme that unites the information in that cluster. Therefore, the expert's memory can be thought of as a pyramid, the top of which contains fundamental principles and concepts, followed by ancillary concepts, with domain-related factual information occupying the lower levels of the pyramid. An expert, or a person who "knows more," has more conceptual clusters in memory, more relations or features defining each cluster, more interrelations among the clusters, and more effective methods for retrieving related clusters than does a novice (Chi and Glaser 1981).

The novice's organizational clusters are structured quite differently from the expert's. In place of the hierarchical arrangement of clusters, the novice's knowledge base is arranged more amorphously, with less attention given to the structural connections, or "pecking order," of the underlying themes that make up each cluster.

Furthermore, the themes themselves are less often arranged in terms of fundamental principles.

A simple yet clever experiment may be used as an illustration (Chi, Feltovich, and Glaser 1981). Expert physicists and novice physics students were given a stack of index cards, each containing a physics problem, and asked to sort the cards into piles according to the similarity in anticipated solutions—that is, problems that could be solved with similar strategies were to be placed in the same pile. Participants in this experiment were told not to solve the problems before sorting the cards. The organizing themes of the piles made by novices were governed by the terminology of the problems. Novices placed problems containing inclined planes into one pile, problems containing pulleys into another pile, and so on. These descriptors are referred to as the problems' "surface features" because they are readily apparent. However, this organizational scheme is considered primitive because it does not help in deciding what principle must be applied to solve the problem. For example, two inclined plane problems could be solved by applying completely different principles and procedures. In contrast to the strategy adopted by the novices, the categories constructed by experts were governed by the underlying principle that could be applied to solve the problem; this is referred to as cuing on the problems' "deep structure." Other experiments indicate that as the novice develops more expertise, the amorphous, surface feature-oriented organizational scheme begins to be restructured and becomes the expert's hierarchical, principle-based organizational scheme.

Experts and novices also differ in the approach they use to solve problems. When given a problem to solve, experts begin by qualitatively analyzing the problem in an attempt to fully understand what it is asking. This qualitative analysis could include: building different representations of the problem to assess which might be the most appropriate; breaking up the problem into easier subproblems; and searching for any principle that could be applied to solve it. Only after convincing themselves that they understand the problem qualitatively do experts begin to quantify their solution. Often quantification is considered "problem solving" in science, when, in fact, much of the essential work takes place during the earlier stage of qualitative analysis. However, because the qualitative analysis phase of the expert's problem solving is not obvious, it often is not perceived, and perhaps worse, it is not taught as an essential part of problem solving. Teachers seldom "brainstorm" a problem in front of students before writing down symbols or equations. Instead students are usually shown the end product, a refined solution, rather than the full process by which the problem was actually solved. Resnick (1983, 478) has commented that: "Extensive qualitative analysis is not common in science or mathematics teaching. It may seem to take too much classroom time, and many teachers are perhaps too inexperienced in these ways of thinking. But the new evidence about learning makes it clear that we cannot avoid taking on this task."

Only during the last 20 years have we come to appreciate fully the importance of qualitative knowledge. Recent attempts to build expert systems in computers have now shown the limits of a purely formal knowledge base. The mathematical structure of scientific theory does not contain the information necessary to decide when and how theory should be applied. Effective application of any theory is impossible without a complex qualitative understanding of the contextual interrelationships. For example, a useful step in a classical Newtonian analysis is to draw a free-body diagram; to do that, one must draw all the forces acting on the particular body. But the formal theory is little help in determining the boundaries of that body. What constitutes an isolated body is an essential, nontrivial aspect of a free-body diagram (the unpredictability of this situation makes it nearly impossible for a computer to handle). Moreover, this aspect is virtually ignored by our curriculum.

Recent work in the history of science also shows the importance of qualitative knowledge in the development of scientific theory. For example, progress in quantum theory was impeded in the early years because physicists did not believe Einstein's radical hypothesis that light was quantized, despite the fact that Einstein was a leading physicist of the period (Szamosi 1986). Although de Broglie's wave-particle hypothesis was objectionable on qualitative grounds to physicists at the time, its rapid acceptance can largely be attributed to Einstein's endorsement of it (Clark 1972). Advances in mathematics and science do not follow a straightforward path, as we would expect if they were the result of a systematic formal process, but instead weave a tangled web wherein many strands are left hanging. It is only in retrospect, when the bulk of the effort has been forgotten, that there appears to have been a systematic investigation (Feyerabend, 1978). Years later some of the hanging strands and lost hypotheses may be rediscovered by students who, with assistance from teachers, can then recognize that even experts have been subject to misconceptions. Thus our lack of emphasis on qualitative approaches to science often shields us from recognizing our student's most important discoveries.

The Teacher as Facilitator/Mediator

In this section, we will discuss what science teachers can do to help students learn science and to retain this knowledge. Because the cognitive research reviewed above is fairly recent, many educators are unaware of it. As a result, science as it is now taught does not reflect the current views of student as learner or teacher as facilitator/mediator.

Learners (students in this context) construct their understanding. New concepts acquire meaning as they are compared with and contrasted to existing concepts. For example, in our first attempt to construct the concept of a whale, we might classify it as a large fish, rather than as a mammal. This is because we think of fish as living in the sea and of mammals as having legs and living on land. Later we learn that the whale is a mammal that acts like a fish in some ways. In this instance we quite literally build a new concept within the context of existing concepts. This is a process that is often repeated and requires the learner's active participation. Learners tie pieces of the new information to existing structures of knowledge. The process helps them to interpret and store new information for easy access in memory. However, if the learner's existing knowledge structure contains naive theories, it is likely that new science experiments and demonstrations will be interpreted in terms of these beliefs rather than in terms of established scientific theories.

There are several reasons that so many students retain misconceptions, even after they have been exposed to standard scientific concepts. First, scientific theories are usually presented quickly and abstractly. If a theory or concept conflicts with the student's beliefs, there is little opportunity for the student to construct an appropriate understanding of the subject matter. Second, teaching as it is practiced today rarely acknowledges common alternative conceptions. It is difficult for teachers to recognize the adverse effect that misconceptions have on learning and the mental effort required by the student to revise them. Third, even when they are aware that misconceptions interfere with learning, teachers have few tools in their arsenal for helping students overcome naive ideas. Teachers are often reduced to paraphrasing the lesson, speaking more slowly and in a louder voice.

Even experienced teachers may fail to construct an accurate model of students' understanding. In a recent study, elementary school teachers with at least 10 years experience tutored six students on difficulties in whole-number addition. Instead of ascertaining what the students knew before attempting remedial instruction, these teachers used a standard classroom lesson that involved a well-defined set of skills and concepts along with activities for teaching and learning those skills. Most of the teachers used a curriculum script normally employed with classes of 20–30 students (Putnam 1987).

Clearly there are teachers who diagnose student difficulties and tailor their instruction to meet students' needs. Yet, this study indicates there still is a need to train seasoned teachers in techniques for diagnosing the obstacles to learning. Instruction should become more bidirectional. Teachers should not only present material but also ascertain whether it is understood. In cases where misconceptions are uncovered, simply telling students that they are wrong and providing a correct explanation is not sufficient. Several new, innovative approaches aimed at helping students overcome misconceptions are beginning to emerge (Clement 1987), but let us describe just one approach that has general applicability (Lochhead and Mestre 1988).

The key to this approach is to challenge misconceptions by probing for conceptual understanding during the course of instruction. A two-pronged effort consisting of asking students questions during class time and analyzing students' answers in quizzes and tests can be used toward this aim. In those cases where the source of the student's misunderstanding is not clear, additional information can be obtained by asking the student to explain how a specific answer was determined. After identifying a misconception, the teacher can then guide the student in the process of dismantling the incorrect notions and accommodating an understanding of the appropriate (or correct) concept. To do so the teacher seeks to create conflict between the student's misconception and the accepted scientific theory by asking some judicious questions that elicit a contradiction, or some obvious inconsistency between the student's beliefs and accepted scientific theory. In confronting the contradiction, the student begins to reconstruct knowledge. The teacher now plays the dual roles of guide to help the student through the process of reconstruction and of monitor to ensure that the student arrives at the desired goal. Finally, it is important to probe the student's understanding sometime later to make sure that the misconception actually has been revised.

The reader might see a paradox in the instructional modifications we are advocating. If misconceptions are as rampant and resilient to change as we have described, then teachers' attempts to address them may be so time-consuming that there would be little time left to cover the requisite subject matter. Luckily, this is not the case. The number of misconceptions that students possess on any given topic is quite small (Resnick 1983; Mestre 1987). Therefore, it is possible to diagnose and to treat misconceptions without any major restructuring of current instructional practices. The benefits are clear. Helping students to tear down incorrect theories and to rebuild them as scientifically acceptable theories will ensure a solid conceptual foundation upon which students can understand the process of theory building as it in fact is practiced by scientists.

In addition, we can help students to build mental structures for storing knowledge efficiently and to adopt techniques for attacking problems efficiently. Because experts possess efficient mental structures for storing knowledge and solving problems, some studies have investigated the effect of exposing novices to similar mental structures. These studies note that novices do exhibit shifts toward "expert" behavior as a result of being exposed to "expert" mental structures.

For example, one study (Eylon and Reif 1984) compared the effect of presenting physics knowledge in a hierarchical fashion with more traditional modes of presentation. Those who received their instruction with the information arranged hierarchically performed significantly better in both recall and problem-solving tasks. Those benefiting most from the hierarchical presentation were the medium-ability students. The high-ability students performed well under either mode of presentation, while the low-ability students performed poorly under either mode. Thus hierarchical presentations of information appear to facilitate retention and problem-solving performance of medium-ability students; high-ability students appear to learn somewhat independently of the mode of presentation. Other researchers (Doyle 1983) have found similar results. The difference in response between medium- and low-ability students may present a dilemma for designers of equal-opportunity programs. Low-ability students may need hierarchically based instruction, but initially they may respond poorly to this style of instruction. Most likely what is needed is a long-term immersion in hierarchically based instruction in an environment that even tolerates and expects an initial period of failure. This approach seems to be behind the success of programs such as Fuerstein's Instrumental Enrichment and the Xavier University SOAR program. (Both of these programs are described in Nickerson, Perkins, and Smith 1985.)

In another study (Heller and Reif 1984) novices were trained to come up with qualitative analyses of physics problems before generating equation-based formal solutions. The analyses required novices to describe problems in terms of concepts, principles, and heuristics. As a result, students were better able to construct solutions to problems. A third study (Mestre and Touger 1989) constrained physics novices to follow a hierarchical problem-solving approach in which they first applied general principles and procedures before they were allowed to use specific equations. Findings revealed a shift toward more behavior typical of the experts' approach to problem solving. After a short period of time, these students increasingly used general principles in categorizing problems and in generating qualitative explanations of physical phenomena.

These studies consistently indicate the benefits that can accrue from instructional techniques that mirror techniques used by experts. A hierarchical structure appears to benefit the presentation of material. More specifically, presentations should begin with the "big ideas" and an effort should be made to show how ancillary ideas, equations, and facts are subsumed by the big ideas. Such an approach helps the student build a hierarchical, richly interconnected network in memory that is efficient for the recall and use of the essential content in the various science disciplines.

Thus, in problem-solving tasks, studies of cognitive processes suggest several modifications in the way science is taught:

- Encourage students to cue on the principles and concepts that could be applied to solve problems and discourage them from cuing on the surface features of problems.
- Encourage students to begin problem solving by using qualitative analyses and discourage them from plunging into quantitative solutions.
- Encourage the development of problem-solving strategies that are applicable across a wide range of situations in different disciplines instead of teaching "prescriptive" approaches to problem solving (for example, "today we will learn about solving inclined plane problems").

This approach should help students make sense out of what they learn. Many students forget facts, equations, and procedures for solving problems shortly after finishing a science course because the information is not stored in an accessible and meaningful way. Science facts, equations, and procedures stored in isolation do not allow the student to apply information to situations different from those in which the information was first learned. Information stored in clusters governed by concepts and principles is retained much longer and is available for application to novel situations.

Teaching the Process of Science: Laboratory and Mathematical Skills

For more than a century, educators have believed that laboratory work is an indispensable part of science education. For example, the educational philosophy behind the founding of Massachusetts Institute of Technology was one of individual discovery through experimentation (Phillips 1981). This philosophy has trickled down to the high school level. The rise of the high school physics laboratory was an attempt not only to conform to the emphasis given to laboratory work in college but also to satisfy college admissions criteria and to popularize the progressive idea that a curriculum based on science was as important as a curriculum based on the classics (Rosen 1954). This emphasis gave rise to the "Harvard 40," a set of experiments "required" of students seeking admission to Harvard University.

It is widely assumed that laboratory work plays three major roles, namely, supporting the lecture portion of science courses, teaching laboratory skills, and teaching the process of science (also called the experimental paradigm). There is little research evidence to support the first belief. For example, the problem-solving ability of two groups of college freshmen was compared in terms of material covered in the lecture portion of a course. One group had conventional laboratory instruction while the same experiments were demonstrated to the second group. The study revealed equivalent performance in problem solving for the two groups when they were tested (Kruglak 1952). A follow-up study (Kruglak 1953) added a third group of students who received no laboratory exposure whatsoever; the findings of this study again revealed no differences in performance among the three groups on a test covering the material presented in class. Studies of chemistry (Dubravcik 1979) and biology (Robinson 1979) students showed similar results when laboratory and nonlaboratory instruction was compared.

Perhaps the greatest danger in using laboratory work to reinforce or demonstrate concepts covered in class is that students get the impression that laboratory work is meant to verify (or even prove) known phenomena, rather than to pose and answer scientific questions (Robinson 1979). Evidence that we convey this attitude to students can be seen in lab reports that often begin by stating, "The purpose of this lab was to prove. . . ."

However, laboratory work is of paramount importance in teaching both skills that are useful in the laboratory and the process of scientific inquiry. The distinction between these two functions should be made clear. Laboratory skills include observation, measurement, laboratory procedure, facility and safety with equipment, and data analysis (both graphing and interpolation). "The process of scientific inquiry" is the ability to pose meaningful scientific questions and to devise a procedure for answering them, the ability to interpret data, make inferences, draw conclusions, and construct models and theories to explain related phenomena.

Research studies indicate that hands-on activities are essential for teaching laboratory skills. For example, in the study cited earlier (Kruglak 1952), the lab group outperformed the demonstration group in manipulative skills and in solving simple problems that required use of laboratory apparatus. In chemistry, training in skills such as weighing and measuring liquid volumes resulted in significant improvements in accuracy and precision (Beasly 1985). Another series of studies found that a laboratory setting was quite effective for teaching laboratory skills to underprepared minority undergraduates majoring in the sciences (McDermott, Piternick, and Rosenquist 1980).

Despite the success of laboratory work for teaching various specific scientific skills, we wish to stress two points. First, we should strive to teach students generalizable laboratory skills. Research findings indicate that students do not retain particular details about specific experiments. The study from which these findings emerged (Brown 1958) revealed that students who had entered a prestigious technical university intending to major in scientific fields did not retain many details concerning the equipment, purpose, and procedures used in specific experiments conducted in high school. However, students in this survey stated that they had selected a scientific career because of their stimulating experience in high school physical science courses. From this, we can conclude that students' laboratory experience should focus on providing intellectual stimulation, on imparting generalizable experimental skills, and on teaching basic experimental design.

Second, laboratory work may not be the most efficient way to teach some skills. For example, it might be more efficient and cost-effective to teach error analysis if students were provided with data rather than requiring students to collect their own data; this approach would alleviate the cognitive overload of having to learn several skills simultaneously while actually doing an experiment. Arming students with the basic skills needed to perform an experiment and to analyze the data increases the likelihood that they will pay attention to the more important experimental issues. If the implicit purpose of laboratory work is to teach the process of science, then teaching lower-level, albeit complicated, skills will likely detract from the students' ability to focus on learning the higher-level skills necessary for the experimental paradigm.

To us the most important function of laboratory work is teaching the *process* of science. Unfortunately, at present, the most common approach to school lab work is often the "traditional," "structured," or "cookbook" lab. Such labs are not optimal for teaching the process of science for several reasons. First, cookbook labs spell out everything for the student including the purpose of the lab, the apparatus, the steps to be carried out, the questions to be addressed, how the data are to be analyzed, and what is expected in terms of a reasonable conclusion. Spelling everything out encourages mental passivity (Tinnesand and Chan 1987). In cookbook labs, the student is seldom, if ever, asked to consider why the questions or hypotheses under consideration are meaningful and why or how the particular apparatus and procedural steps were selected. In short, cookbook labs do nothing to exercise the mental muscles that a scientist must flex in posing meaningful questions and in devising experimental techniques for answering those questions.

At the other extreme are "open-ended" labs where students pose their own questions or hypotheses, design an experimental approach to answer the questions, build or find apparatus for conducting the experiment, analyze their findings, and draw the appropriate conclusions. Open-ended labs are very time-consuming and virtually impossible to carry out in standard high school sessions involving 20–30 students.

We propose a compromise between these two approaches—an approach we call "structured inquiry." The idea is to provide a structure within which to carry out experimental work, but one that allows students the freedom to explore and learn on their own. Rather than providing all the details of an experiment, as in cookbook labs, the structured inquiry approach provides students with both a designated topic that will be the focus of the experiment and the equipment to conduct the experimental work. Within this framework, the teacher acts as a mediator.

Laboratory activities would begin with a classroom discussion of experimental design. The goal of this discussion is to help students pose questions or hypotheses about a topic that can be answered using the equipment available and to help students devise an experimental procedure by which the questions or hypotheses can be addressed. At first, posing meaningful questions and devising experimental techniques for answering them may be difficult for students, so the teacher must be ready to provide necessary assistance until students are capable of completing these two steps on their own. At the learning stage, wide-ranging classroom discussion is likely to generate several experimental questions and procedures, all of which cannot be investigated by an individual student. To ensure exploration of all relevant questions, the teacher can divide the class into collaborative groups, each of which can explore a particular question or related set of questions. Once this classroom discussion is completed, the students carry out the experimental work. After the students have analyzed the data and drawn preliminary conclusions, the class can reconvene for another discussion, this time about the findings and conclusions regarding the questions and hypotheses posed. This discussion provides the collaborative groups with an opportunity to pool their results and the teacher with an opportunity to serve as mediator in a discussion of the experimental findings. Unlike cookbook labs, this approach encourages students to be active participants in experimental design.

Learning Outcomes and the Structured Inquiry Approach

In the last section of this chapter, we will describe the learning outcomes that *all* students should achieve in high school as first set forth in *Academic Preparation for College: What Students Need To Know and Be Able To Do* (College Board 1983). In addition, we will draw on two simple elementary, physical science experiments in order to make our discussion of the desired outcomes and the structured inquiry approach more concrete and meaningful. We begin by discussing the two experiments in some detail to streamline the subsequent discussion of both the outcomes and the structured inquiry approach.

The experiments explore the cooling–heating curve of two substances: water and paradichlorobenzene.* The cooling–heating curve of a substance describes the functional relationship between temperature and time as heat enters or leaves the substance. The study of such curves is an ideal subject for eliciting students' preconceptions concerning this subject. If asked to state what would happen if a hot substance is placed in cool surroundings, nearly all students will state that the substance will cool off, meaning that heat energy will be exchanged from the hot substance to the surroundings. If asked what would happen to the temperature of the hot substance as time passes, students will further state that the temperature will decrease until the substance and its surroundings reach equilibrium. The same arguments carry over to a cold substance placed in warm surroundings, only now the substance is warming, its temperature rising as a function of time until equilibrium is reached.

These commonsense preconceptions are quite valid as long as the substance under consideration does not undergo a phase change (for example, from solid to liquid). When a substance undergoes a phase change during heating or cooling, its temperature will remain fairly constant until the phase change is completed. This situation is illustrated for water and paradichlorobenzene in Figures 1 and 2, which show a heating curve for water and a cooling curve for paradichlorobenzene, respectively.

To obtain the heating curve of water shown in Figure 1, we start with a piece of ice at −25°C and heat it uniformly. If we take temperature readings as time passes (starting with a thermometer frozen into the ice), we would observe the behavior shown on the portion of the curve labeled "ice." This section shows the ice heating up from −25° C to 0° C; note that only the solid phase is present during this section. When the ice reaches 0° C, it begins to melt. Now we see that this part of the curve, labeled "ice and water," flattens out, which indicates that a phase change is taking place. In this case the change is a solid-to-liquid transition (ice turning

*Paradichlorobenzene—chemical formula $C_6H_4Cl_2$—is a simple chemical available in hardware or drug stores under the name "moth ice," "moth crystals," or simply "mothballs."

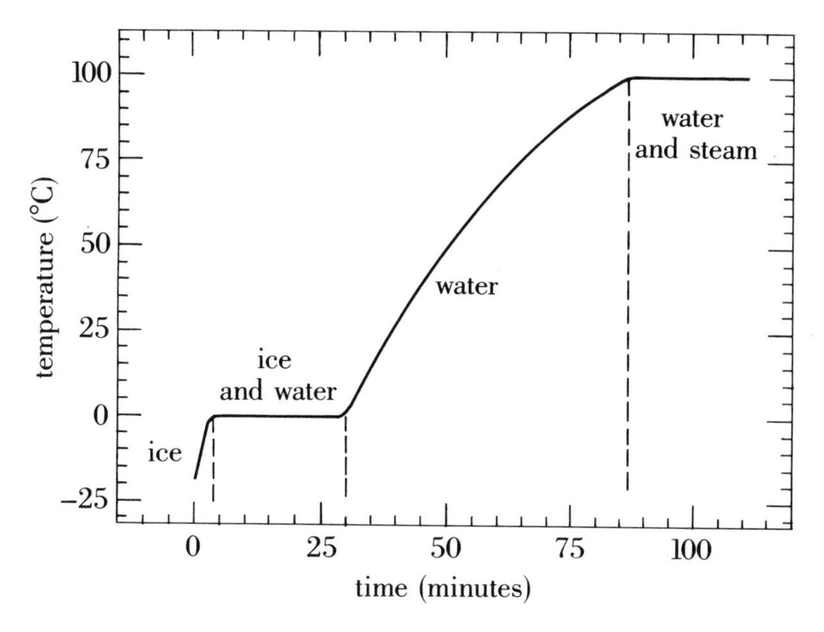

Figure 1. Heating curve for water (H_2O).

to water). The temperature during the phase transition is counter-intuitive to most students' notion of a heating process because although the ice and water are being heated, the temperature does not rise. As implied by the flatness of the curve, the temperature remains constant as time passes. In fact, the temperature remains constant at 0° C until all of the ice melts. The temperature remains constant because 334.7 joules of heat are needed to melt each gram of ice at 0° C to water at 0° C. The "latent heat of fusion" for water is 334.7 joules per gram; this is the amount of heat needed for a phase transition of one gram of ice at 0° C to one gram of water at 0° C.

After all of the ice melts, the water begins to heat up, and the curve shows a steady rise in temperature. When the temperature of the water reaches 100° C, another phase transition begins to take place. This time, water at 100° C turns into steam at 100° C. The "latent heat of vaporization" of water is 2,259.4 joules per gram. Again we see that although the temperature curve does not rise during the transformation to steam, the water is nonetheless taking in heat.

The situation depicted in the cooling curve of paradichlorobenzene is similar (see Figure 2), but in this case we start with liquid paradichlorobenzene at 100° C (produced by placing a small amount in a test tube that is put in boiling water until the paradichlorobenzene melts, and the liquid reaches 100°C). Cooling is achieved by removing the test tube containing the liquid paradichlorobenzene from the boiling water and letting it stand at room temperature. Again, this cooling curve shows the effect of a phase change. When the liquid paradichlorobenzene reaches its melting point of 53° C, it begins a phase transition to solid paradichlorobenzene. The temperature remains fairly constant at about 53° C until all the liquid paradichlorobenzene solidifies. Again, although the liquid–solid combination of paradichlorobenzene is

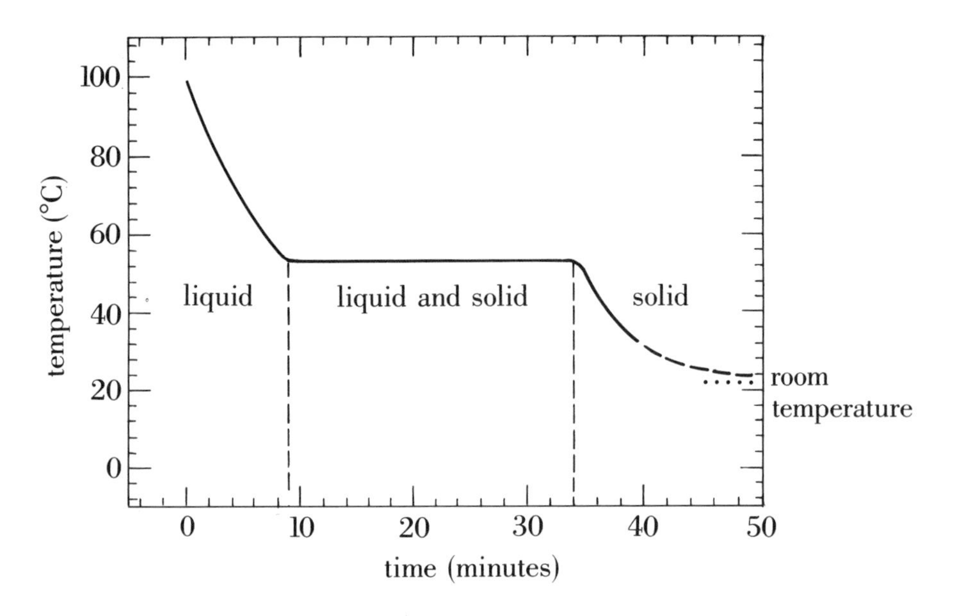

Figure 2. Cooling curve for paradichlorobenzene ($C_6H_4Cl_2$).

actually cooling during the phase transition, the temperature does not decrease until the transition is complete.

Other features of cooling–heating curves can be used to deduce properties of the substance being studied. If the heat flow remains constant (that is, the same number of joules of heat enter or leave the substance each second), then the length of time necessary for a phase transition (that is, how long a cooling or heating curve remains flat) is an indirect measure of the amount of substance present. For example, if the flat portion of a water-ice heating curve for one sample is four minutes, and that for another sample is eight minutes, then the second sample contains about twice the mass of the first sample. This is clearly the case because the length of time needed to convert all of the ice to water is dependent on the number of grams of ice in the sample.

In addition, the slope (steepness) of any portion of a cooling or heating curve is an indirect measure of the "specific heat" of the substance. The specific heat of a substance is a numerical value that determines how much heat is necessary to raise the temperature of one gram of the substance by one degree centigrade. The specific heat of water is 4.18 joules per gram per degree centigrade, meaning that it takes 4.18 joules of heat to raise the temperature of one gram of water by one degree centigrade. If we consider the two rising portions in Figure 1, in which the ice and the water are being heated, it is evident that the slope of the ice portion is steeper than the slope of the water portion. This means that, for a fixed amount of heat intake by the sample, it takes more heat to raise the temperature by some specific number of degrees of a given amount of water than it does to raise the temperature of the same amount of ice. Therefore, the specific heat of ice is lower than the specific heat of water.

The cooling–heating curve experiments are simple to carry out, yet illustrate many aspects of the experimental nature of science. [†]We will now draw on these examples to discuss the science

---

[†]Teachers should note that in practice it can be difficult to find equipment that gives good results. Some thermometers are distorted when frozen and give temperature readings that are off by 10° C or more, so teachers should test equipment before beginning work with students.

learning outcomes set forth in *Academic Preparation for College.* It is important to keep in mind that this example will be used to illusrate all science outcomes; in practice, a teacher may wish to use a particular lesson to focus on a specific outcome.

Approaching Scientific Questions Experimentally

- *Outcome A: Sufficient familiarity with laboratory and fieldwork to ask appropriate scientific questions and to recognize what is involved in experimental approaches to the solution of such questions.*

A.1 Identifying and posing meaningful, answerable scientific questions.
A.2 Formulating working hypotheses.
A.3 Selecting suitable methods for answering scientific questions or testing hypotheses.
A.4 Designing appropriate procedures for laboratory or fieldwork to test hypotheses.

This outcome is one of the most difficult. It is not easy for students to recognize that some important questions cannot be answered scientifically (for example, whether human life begins at conception or at birth). Even after an answerable scientific question has been posed, the task of designing the appropriate experimental techniques and procedures needed to answer the question can be a long, tedious process involving the construction of new apparatus. Nevertheless, this outcome is the cornerstone of scientific inquiry, and there are various ways of helping students to become proficient at attaining it.

The teaching of this outcome through a structured inquiry approach would begin with classroom discussion. The teacher initiates the discussion by presenting the topic for discussion and explaining what equipment is available for experimental work. In the case of the cooling–heating experiments, the teacher would begin by stating that the topic is the study of the heating and cooling behavior of substances. The teacher would then continue by stating that, in theory, any substance can be studied, but that right now only two substances will be considered, water and mothballs. It should be made clear to the students that any conclusions resulting from the study of these two substances can also be made about the heating–cooling behavior of other substances.

Within this framework, the teacher begins by engaging the class in a discussion of heating and cooling. Initially, the aim should be eliciting students' preconceptions about heating and cooling, and in particular, the relationship between temperature and heating or cooling. Two preconceptions, namely that the temperature of a substance increases during heating and decreases during cooling, will likely emerge shortly after the discussion begins. The teacher can then ask if cooling always lowers temperature and, conversely, if heating always raises temperature. There should be further discussion of these ideas. If students only bring up single-phase substances (for example, pure solids or pure liquids), the teacher can introduce phase transitions by asking, "What happens if you start with ice and heat it slowly? Would its temperature rise steadily?" These questions should generate discussion of what it means to melt something and of the behavior of temperature during melting.

After the main issues are brought out and discussed, the teacher can ask students to identify questions that could be answered using the equipment available (A.1). If the initial discussion was fruitful, then a series of meaningful questions should be forthcoming. For example: Will the temperature of a substance always rise when heat is added? Will the temperature of a substance always decrease when heat is extracted from it? Will the temperature of a substance rise while it is melting? Will the temperature of a substance decrease while it is freezing?

Ill-defined questions, or questions that cannot be answered either scientifically or with the available equipment, can be used as a springboard for discussions of what constitutes a scientific, answerable question. For example, a question such as "Exactly how much heat does it take to melt 10 grams of ice?" cannot be answered with the equipment available to the students. The teacher can draw distinctions among questions that cannot be answered through experimentation, questions that are scientifically meaningful but cannot be answered with the equipment available, and ill-defined questions that, when refined, are amenable to scientific inquiry through experimentation.

The next step is to guide students toward understanding the distinction between a meaningful scientific question and a hypothesis (A.2). The goal should be to help students understand that a hypothesis is a statement that draws a general conclusion. Two hypotheses might be: "All substances exhibit a rise in temperature when heated," or "The temperature of a substance does not change during a phase transition." Here, only the second is valid. That a hypothesis has predictive value, and that several scientific questions can be posed to test the validity of a hypothesis, should be stressed.

A class discussion of methods and procedures for answering the scientific questions posed should follow (A.3 and A.4). For specificity, let us assume that the discussion focuses on designing an experimental procedure for answering the question, "Does the temperature of paradichlorobenzene always decrease when it is cooled?" Important issues to consider here are: How is the paradichlorobenzene going to be cooled? What temperature range can be achieved during the cooling period? Is the rate of cooling an important factor to consider? (If cooled too fast, there may be slight temperature variations between the outer regions of the substance, which are in contact with the cold reservoir, and the inner regions, which are not. The cooling–heating rate is particularly crucial in the heating-of-ice portion of Figure 1; if the ice is heated too rapidly, the outer layers will begin to melt while the inner layers remain below 0° C).

After several procedures are designed for attacking the scientific questions posed through experimental work, the teacher can divide the class into suitable groups to work on specific questions. Each group might consist of several pairs of students working together as teams. Now the class is ready for hands-on experimental work. The next outcome can be covered within the context of hands-on laboratory activities.

Gathering Scientific Information

- *Outcome B: The skills to gather scientific information through laboratory, field, and library work.*

B.1 Observing objects and phenomena.

B.2 Describing observations accurately using appropriate language.
B.3 Assembling the appropriate measuring instruments.
B.4 Measuring objects and changes quantitatively.
B.5 Analyzing observational and experimental data.
B.6 Developing sound skills in using common laboratory and field equipment.
B.7 Performing common laboratory techniques with care and safety.

After the students have posed scientific questions and designed procedures for answering them, hands-on laboratory work begins. Unlike cookbook labs where all the details of the procedures are spelled out for the student, the structured inquiry approach allows the student to think about an approach for implementing the procedures proposed during the classroom discussion. The teacher would then walk around the lab and monitor progress, making suggestions and providing hints to hone students' laboratory skills.

For example, the kinds of activities that the teacher should encourage under outcomes B.3 and B.4 include deciding how many samples of paradichlorobenzene or water–ice to use in various subexperiments; weighing and recording the samples in order to observe the effect of mass; devising a clear plan for taking temperature–time readings; deciding how to arrange the thermometer to take accurate readings of the temperature of the sample. (If they start with ice, students should realize that the mercury reservoir of the thermometer should be near the center of the ice sample.)

Under outcomes B.1 and B.2, students should be encouraged to make a data table with appropriate headings and labels to record all quantitative readings of time and temperature. It may be necessary to remind students that observing physical changes in the substance as the temperature varies is as important as making accurate quantitative measurements. They need to observe and record the time intervals during which there is only a solid present, the time intervals during which there is both a solid and a liquid present, the time intervals during which there is only a liquid present, and the time intervals during which there is both a liquid and a gas present.

Students should be encouraged to use safety precautions and sound laboratory skills (B.7). For example, students should be cautioned that an odoriferous substance such as paradichlorobenzene should not be boiled without a hood to exhaust the fumes, because the fumes could be noxious or even toxic. Other common safety practices include wearing goggles and handling hot objects with caution.

Teachers can encourage the development of sound skills in using common laboratory equipment (outcome B.6) while the experiments are being performed. For example, students should stir the contents of the sample (when there is a liquid, or both a liquid and a solid present) to ensure uniform temperature readings. Students should be admonished not to hold the samples while collecting data; contact with the hand could provide the sample with a heat source or a heat sink, depending on the temperature of the sample. The teacher also could introduce the important experimental concept of reproducibility by asking, "What guarantee do you have that your data are accurate?" Students should be led to the realization that it is sound experimental practice to perform an experiment at least twice. The teacher's questions can also be a good lead-in for a discussion of the difference between accuracy and precision; in particular, students should learn the concept of *significant figures* in taking measurements.

After the experiments are completed, observational and experimental data are analyzed (B.5). Here, students can be encouraged to plot their data, making sure that all relevant information is appropriately labeled (that is, title for the graph, labels for the axes, the mass of the sample, clear demarcations of the solid-only, solid-liquid, and liquid-only portions of the plot). After this step is completed, students are ready to proceed to the next two outcomes, namely the ability to organize and communicate results and the ability to draw conclusions.

Organizing and Communicating Results

- *Outcome C: The ability to organize and communicate the results obtained through observation and experimentation.*

C.1 Organizing data and observations.
C.2 Presenting data in the form of functional relationships.
C.3 Extrapolating functional relationships beyond actual observations, when warranted, and interpolating between observations.

Now that students have tables and graphs displaying the cooling–heating curves, they can begin to organize their findings and observations. For example, students should be guided toward discerning that the mass of the sample has an effect on both the flat and the sloping portions of the graphs; for a constant rate of heat flow, the larger the mass, the longer the flat portion of the graphs and the less steep the sloped portion of the graphs (explanations for these phenomena will be discussed in the next section on drawing conclusions). The results should make students realize that phase transitions have a clear effect on temperature during cooling or heating processes, and in particular that during a phase transition the temperature of the sample remains fairly constant. Finally another, perhaps more subtle, feature of heating and cooling curves is that the slopes of the all-solid and all-liquid portions are different.

Writing functional relationships (C.2) that describe data is often a very difficult task for students. In the case of the cooling–heating experiments, the teacher can help students deduce two functional relationships. The first is fairly straightforward and has to do with the behavior of temperature during a phase change. The teacher could begin by asking students to describe verbally the temporal behavior of temperature during a phase change. Since the students' plots display that temperature remains relatively constant during a phase change, guiding students to write down the appropriate mathematical relationship for temperature should not be difficult. This relationship is simply T = constant, where T denotes the value of temperature. The teacher could then proceed to ask for the meaning of the constant. To be able to attribute meaning to the constant, students need to perceive that both solid and liquid are present during a phase change and thereby deduce that the constant must be the temperature at which the substance melts (or freezes).

To help students deduce the second, more difficult functional relationship, the teacher may need to help students apply knowledge acquired in algebra. This second relationship has to do with the behavior of the temperature for the single-phase portions of the cooling curves. Here the teacher may need to begin by helping students recall the equation for a straight line. This done, students can apply the equation to describe the temporal behavior of the temperature during the single-phase portion of the graph. In the case of cooling experiments, the goal is to guide students to the functional relationship, $T = -at + b$ (or $T = at + b$ for heating experiments), where T is the temperature, t is time, a is the slope of the line, and b is the value where the line crosses the temperature axis (in this instance, there is no particular meaning attached to b, the *y-intercept*, since the graphs are made up of different parts denoting different physical pheonomena).

After these functional relationships are obtained, students can be encouraged to predict the salient features of cooling or heating curves for samples they did not investigate in the laboratory by extrapolating from their data (C.3). One question that a teacher might ask to promote discussion is, "How would the flat region of the cooling–heating curve for an 8-gram sample of paradichlorobenzene undergoing a liquid–solid phase transition differ from that of a 4-gram sample?" Those students who understand the role of mass in the cooling–heating curves will be able to deduce that the required time for the phase transition for the 8-gram sample will be twice as long as that for the 4-gram sample. Perhaps the biggest challenge for the teacher is helping students deduce that the slopes of the single-phase portions of the curves are also proportional for the amount of mass in the sample. It is important that students realize that these two conclusions hold only if heat leaves or enters the samples at the same rate.

Drawing Conclusions

- *Outcome D. The ability to draw conclusions or make inferences from data, observation, and experimentation, and to apply mathematical relationships to scientific problems.*

D.1 The ability to interpret data presented in tables and graphs.
D.2 The ability to interpret, in nonmathematical language, the relationships presented in mathematical form.
D.3 Evaluating a hypothesis in view of observations and experimental data.
D.4 Formulating appropriate generalizations, laws, or principles warranted by the relationships found.

The most interesting, and subtle, scientific concept in the cooling–heating curve experiments is the temperature–time behavior during a phase transition. The graphs made during the experiments should clearly indicate that the temperature remains constant during a phase transition, and students should now be able to evaluate this finding by reconsidering any hypotheses made prior to performing the experiments (D.3 and D.4). For example, if the hypothesis, "all substances exhibit a rise in temperature when heated," is examined, students will now be able to refute that hypothesis with their own data and observations. More important than refuting the initial hypothesis is the ability to come up with different hypotheses that could explain the data and observations. The teacher can help students to realize that this hypothesis is correct as long as the substance is not undergoing a phase transition. Thus, the hypothesis can be made consistent with all experimental findings by a simple modification: "All substances exhibit a rise in temperature when heated, unless a phase transition is taking place in which case the temperature will remain constant until the phase transition is completed."

It is also important that students modify their preconceptions that the temperature of a substance must rise during heating and drop during cooling. Their data should illustrate that cooling (heating) takes place during phase transitions without a drop (rise) in temperature. This finding can be interpreted (D.2) to mean that heat must flow in or out of a substance during phase transitions. To solidify this concept, teachers can relate these findings to students' own life experiences. In particular, all students know that one way to cool a drink is to put ice in it. They should now be able to offer a scientific explanation for this common practice: putting ice in a drink cools it because a substantial amount of heat is needed to melt ice, and thus every gram of ice that melts takes heat from the drink, thereby lowering the drink's temperature. Students should then also realize why the practice of pouring water at 0° C into drinks has not caught on as a method for cooling. Because no heat is used to melt ice, it would take substantially more water at 0° C to cool a given amount of liquid to the desired temperature and would make the drink very dilute.

Students can also be assisted in interpreting the mathematical relationships (D.1 and D.2) found during the analysis of their data, as well as in generalizing the mathematical relationships (D.4). For example, it is straightforward to interpret the interdependence of time and mass for the flat portions of cooling–heating curves (that is, the more massive the sample, the longer it takes to complete the phase transition). The scientific reasoning explaining this phenomenon suggests that a certain amount of heat is needed to melt each gram of ice. Assigning the variable l (for latent heat) to this amount of heat, then to melt m grams of ice would take lm heat energy. The total heat (lm) needed to melt the ice is not supplied instantaneously; it flows at some constant rate, depending on the method used to heat the ice. Suppose it takes a certain amount of time to melt the sample of ice, t_{melt}. An equation is not changed by multiplying by t_{melt}/t_{melt}, so we have

$$(\text{Total heat needed to melt } m \text{ grams of ice}) \times (t_{melt}/t_{melt}) = lm$$

or

$$(\text{Total heat needed to melt } m \text{ grams of ice}/t_{melt}) \times (t_{melt}) = lm$$

so

$$t_{melt} = lm/(\text{Total heat needed to melt } m \text{ grams of ice}/t_{melt})$$

This last equation shows that the time it takes to complete the phase transition is directly proportional to the amount of ice in the sample, m, directly proportional to the amount of heat needed to melt one gram of ice, l (which is a property of ice), and inversely proportional to the rate (the quantity in parentheses) at which heat is flowing into the sample. Thus, this equation clearly demonstrates that doubling the mass of a sample doubles the time it takes for the phase transition, or that increasing the rate of heat flow decreases the amount of time it takes for the phase transition. These ideas can, of course, be extended to the phase transition of paradichlorobenzene, or to liquid-gas phase transitions.

A similar procedure can be used to interpret and generalize the findings of the sloping portions of cooling–heating curves. Here, the teacher could argue that it takes a certain amount of heat to raise the temperature of one gram of substance by one degree centigrade. (Similarly, it takes the same amount of heat to flow out of a substance to lower the temperature of one gram by one degree centigrade). This value is just the *specific heat* of the substance (usually denoted by the variable symbol c). The total amount of heat needed to raise (or lower) the temperature of m grams of the substance from temperature T_1 to temperature T_2 is:

Heat to raise the temperature of m grams from T_1 to temperature $T_2 = mc\ (T_2 - T_1)$.

Dividing both sides by the time interval it took to effect this rise in temperature, $t_2 - t_1$, gives:

(Heat to raise the temperature of m grams from T_1 to temperature $T_2)/(t_2 - t_1) = mc\ (T_2 - T_1)/(t_2 - t_1)$.

Solving for the slope of the graph of temperature versus time yields:

Slope $= (T_2 - T_1)/(t_2 - t_1)$
$=$ [(Heat to raise the temperature of m grams from T_1 to temperature $T_2)/(t_2 - t_1)] \times [1/mc]$

This last equation makes the slope's behavior understandable. The slope is directly proportional to the rate of heat flow (increasing the rate of heat flow increases the slope) and inversely proportional to the mass of the sample and the specific heat of the substance, which both decrease the slope if these quantities are increased. This equation explains why the slope of the ice portion of Figure 1 is steeper than that of the liquid portion: the specific heat of ice (2.09 joules per degree centigrade per gram) is lower than that of water (4.18 joules per degree centigrade per gram). This equation also explains, for a constant heat flow rate, why the steepness of the slope decreases as the mass of the sample is increased.

Clearly, some complex reasoning on the part of students is necessary to attain this learning outcome. In particular, students must draw on their mathematical knowledge to make sense of the experimental findings. The majority of students may not be able to carry out the steps leading to this outcome on their own, so the teacher must carry the burden of guiding the students through the reasoning necessary to draw meaningful conclusions from the experimental data. In the structured inquiry approach, this outcome is best taught within the setting of a classroom discussion in which students can pool their individual experimental findings and collective brain power.

Recognizing the Role of Experimental Work in Constructing Theories

- *Outcome E. The ability to recognize the role of observation and experimentation in the development of scientific theories.*

E.1 Recognizing the need for a theory to relate different phenomena and empirical laws or principles.
E.2 Formulating a theory to accommodate known phenomena and principles.
E.3 Specifying phenomena and principles that are satisfied or explained by a theory.
E.4 Deducing new hypotheses from a theory and directing experimental work to test it.
E.5 Formulating, when warranted by new experimental findings, a revised, refined, or extended theory.

The previous outcomes have illustrated the experimental paradigm in terms of selecting a topic for investigation, devising meaningful (and answerable) scientific questions, designing experimental procedures for answering these questions, and finally analyzing the data and arriving at a synthesis so that appropriate conclusions can be drawn. If we stopped here, science would merely be a compendium of observations and conclusions, without any underlying, overarching explanations to tie together related phenomena. A scientific theory explicates and unifies a body of scientific knowledge in a consistent and rational manner. Unlike a hypothesis, which attempts to predict a single phenomenon, a scientific theory not only predicts related phenomena but also suggests new directions that may generate new hypotheses and experiments that ultimately serve to extend or refine the theory and, hence, the frontiers of scientific inquiry.

Any systematic set of experiments, such as the cooling–heating experiments we have discussed, should convince students of the usefulness of scientific theories for explaining related phenomena (E.1). Formulating a theory (E.2) is much more difficult. However, armed with the experimental findings of the cooling–heating experiments and with the assistance of the teacher, students can begin to formulate a theory that explains and extends the cooling–heating phenomena they observed in the laboratory. For example, the following might be postulates of this theory:

1. Heat is a form of energy.
2. Temperature is a measure of how much heat energy an object possesses. The temperature of a substance is an aggregate measure of the energy of motion possessed by the molecules of that substance. The faster an average molecule is moving in the substance, the higher is its temperature.
3. Heat energy is exchanged between bodies at different temperatures that are within close proximity of each other.
4. A unique property of every substance in a pure state is a number characterizing the amount of heat energy needed to raise or lower the temperature of one gram by one degree centigrade. (This is the *specific heat* of the substance.)
5. A unique property of every substance undergoing a phase change is a number characterizing the amount of heat energy needed to change the phase of one gram of substance at its melting (or evaporating) temperature. (This value is the *latent heat* of the substance.)
6. When a substance is undergoing a phase transition, its temperature remains constant until the phase transition is complete and the substance is once again in a pure state. Heat energy is being exchanged during phase transitions, but this energy goes into the melting (freezing) process, or into the evaporating (condensing) process.

The postulates provided by students will, of course, not be as precise or elegant as these. However, the students' postulates will likely touch upon the content of the postulates listed here. What is important to note is that they not only explain the experimental findings of the cooling–heating experiments but also go beyond the experimental findings by suggesting new hypotheses to test the theory. For example, postulates 4, 5, and 6 are all that is needed to explain the behavior of cooling–heating curves (E.3). Postulate 4 explains why the temperature of a substance in a pure state rises (falls) when heat energy enters (leaves) it, why the slopes of the cooling–heating curves of the same amount of any two different substances in the same state (for example, liquid paradichlorobenzene and water) are different for a constant rate of heat exchange, and why the slopes of the cooling (heating) curves of different pure states (for example, ice and water) for the same amount of different substances are different. Similarly, postulates 5 and 6 explain why, for a given rate of heat exchange, the length of time needed to complete a phase transition is proportional to the amount of substance present; why, for a given rate of heat exchange, different lengths of time are needed to complete a phase transition of the same amount of different substances (for example, melting 10 grams of paradichlorobenzene and melting 10 grams of ice); and why cooling or heating takes place during phase transitions without any change in temperature.

This heat theory also suggests new hypotheses to test the theory (E.4). For example, postulates 4 and 5 imply that cooling and heating are reversible processes. More specifically, postulate 4 implies that it should not take any more or less heat to cool 10 grams of water by 20 degrees than to heat 10 grams of water by 20 degrees. Two simple experiments can be performed to test this hypothesis. The first is to mix two equal amounts of water at two different temperatures. If the final temperature of the mixed batch is the average of the two initial temperatures, then the hypothesis is supported. In a second experiment, two unequal amounts of water at different temperatures are mixed; the final temperature of the mixed batch should be a predictable quantity. For example, if 40 grams of water at 90° C is mixed with 80 grams of water at 60° C, then the 40 grams of water should go down in temperature twice as much as the 80 grams of water goes up in temperature. This would lead to a prediction of 70° C as the final temperature of the 120 grams of water. Students can also be encouraged to derive the mathematical relationship that predicts the final temperature of two samples, m_1 at temperature T_1, and m_2 at temperature T_2 (the relationship is $T_{final} = [m_1T_1 + m_2T_2]/[m_1 + m_2]$).

Certain experiments designed to test the theory may require modifications or revisions in the postulates (E.5). For example, if heating some particular substance results in a chemical reaction, then postulates 4 and 5 may need some revision or qualification to include physical changes and exclude chemical changes. Another modification may be required for postulate 3. If two bodies at different temperatures are placed within close proximity of each other, but a layer of insulation shields one body from the other, then little heat exchange will take place between the two bodies. This can be demonstrated by placing 50 grams of water at 90° C in a beaker exposed to the air, and another 50 grams of water at 90° C in a well-insulated thermos. Temperature readings will reveal that heat is exchanged between the air and the water in the beaker, but that little heat is exchanged between the water in the thermos and the outside air. Students might also be asked to use one of the postulates, or to devise a new postulate, to explain why touching a piece of metal feels colder than touching a piece of wood or a piece of paper at the same temperature.

Thus far we have not mentioned postulates 1 and 2. These postulates are not a direct result of the experimental findings from the cooling–heating experiments, but rather inferences based on these experiments. They are an attempt to define heat in fundamental terms. Because these postulates are more abstract, it is harder to devise experiments to test them. However, students can be aided in constructing a model that explains the connection between temperature and heat energy. If we think of a substance as composed of molecules in constant motion, then we can argue that the faster the molecules are moving, the more energy they contain. The teacher can make a direct analogy to everyday experience to convince students of this argument. For example, it is intuitively obvious that a car moving at 50 miles per hour has more energy (and can therefore do more damage in a collision) than the same car moving at 10 miles per hour. What remains is to show a connection between the speed of the molecules in a substance and temperature. The teacher might best demonstrate this relationship by having the students clap their hands 20 times under two different conditions, first moving the hands at a slow speed and then moving the hands at a faster speed. The hands will get hotter when they move at a faster speed. Thus, in this model, higher speeds are indirectly measured by high temperatures, while lower speeds are measured by low temperatures.

This molecular model of temperature can also be used to explain how two samples of water reach their final temperature in water-mixing experiments. The average speed of a molecule in the hot batch of water is higher than the average speed of a molecule in the cold water. When the two batches of water are mixed, collisions take place, causing an exchange of energy among all the molecules and yielding an "averaging" of the speeds of the molecules.

III: The Science Curriculum: Goals and Commentary

Teachers may expect a chapter on curriculum to provide a list of topics to be covered in high school science courses and to prescribe the sequence in which these courses should be taught. However, this is not our purpose in the following discussion. Instead, we hope to consider a number of points bearing on current debate about the science curriculum and explore their implications for curricular reform. A variety of course and content lists are readily available, of course, for teachers who want them; and many states have their own specific curriculum requirements. For future reference, teachers will want to watch closely the work of Project 2061 (sponsored by the American Association for the Advancement of Science) in determining what knowledge students should acquire before graduating from high school if they are to be scientifically literate citizens (AAAS 1989).

One reason we wish to avoid our own detailed statement about curriculum content is that, in practice, such statements are often part of the problem in science education. Although content lists can serve as guideposts, several caveats are in order regarding their use. First, by their very nature, such lists emphasize content over cognitive processes. It would be much more balanced if we had process lists to accompany content lists. Second, descriptions of the content of science courses are often regarded as complete and exhaustive. As a result, many believe that as long as all the topics are covered, students' educational needs are met. We do not agree.

Moreover, the sequence of science courses does not have to follow the traditional order, i.e., earth science, biology, chemistry, and physics. Courses can be offered in any order so long as the sequence is carefully thought through. While it is essential that students have the appropriate mathematical background before taking a particular science course, new developments in software for calculators and computers are changing the constraints here as well. It is now possible, for example, to teach material that once required an understanding of differential equations using simulation languages, such as Stella (Mandinach and Thorpe 1987).

This chapter does set out some general goals for the science curriculum. Meeting these goals will make science more accessible, meaningful, and interesting to students, especially to those groups who have tended to avoid science in the past. This is essential if we are to attract more students to professions in the pure and applied sciences. In addition students who do not pursue careers in science will gain the knowledge and skills they will need throughout their lives.

Integrating Content and Process in Science Instruction

The most significant debate in education today focuses on content versus process. Those who argue for content point to the deplorable performance of students in competency assessments, such as the National Assessment of Educational Progress (NAEP) (Dossey et al. 1988), and believe that a back-to-basics approach in which teachers cover more mathematics, science, reading, writing, and social studies will restore the level of knowledge it is thought students once possessed. Content advocates also point to the explosion of knowledge and are correct in stating that we will need to deal with greater amounts of information in the future. Those who argue for process also point to the NAEP results, but indicate that students who do quite well with routine "textbook" problems and factual questions have inordinate difficulties in nonroutine multiple-step problems. They also argue that content without process will turn students into "factual data bases" without increasing what are generally called *higher order thinking skills*.

In point of fact, both sides are right. Students need a broad knowledge base with which to reason but they also need the higher order thinking skills necessary to apply this knowledge in working out complex problems. It is time to reconcile the two sides and stop fueling the debate.

On the content side, there is certainly an information explosion in science. Let us cite but one example. *Chemical Abstracts* summarizes information appearing in major professional journals in the field. Over a period of 32 years, *Chemical Abstracts* published one million entries. In the two years that followed, that number was doubled (Fensham and Kornhouser 1982). The other scientific disciplines have experienced similar growth. Clearly, if we try to include *all* of this information in a science curriculum, we are doomed to failure given the limited storage and processing capacities of the human mind.

Therefore, it is imperative for a student to glean from the scientific disciplines a core body of knowledge that is crucial for a beginner to participate in each discipline. Choosing such a core of knowledge is pivotal. Include too little and the student lacks the necessary base for acquiring more knowledge and solving complex problems; include too much and we overcrowd the curriculum, leaving little time for teaching higher order thinking skills. Bear in mind here the argument presented in the previous chapter that the way knowledge is presented can be as crucial as what the knowledge happens to be in determining students' ability to acquire and retain it.

Teachers should recognize that there is no single best way to select the material to be covered. Nonetheless, core knowledge from a particular discipline must be chosen, and then students must be helped to build a useful hierarchical network within which to store this knowledge. A hierarchical network, no matter how skeletal the framework, will provide students with a structure for handling increased amounts of, or changes in, information later on. We have already discussed how the "big ideas" need to be emphasized and how ancillary ideas and facts have to be tied to these big ideas. Many of these are common to several scientific disciplines. For example, the concept of conservation of energy can be taught equally as well in biology, chemistry, and physics as in marine science, environmental science, and earth science; energy conservation can be illustrated by combustion, collisions, ocean currents, "waste heat," and tidal waves. There is a danger, of course, that nontraditional science courses oriented to students outside the standard college track can be watered down. However, we believe the rigor of the scientific process can and should be taught in any science course. We should not convey the impression that "real" science is biology, chemistry, and physics, while nontraditional science courses are survey courses for students who are afraid of science. We believe that a serious spirit of scientific inquiry can be conveyed at all levels of instruction.

It is important to acknowledge that those who advocate more emphasis on process in the classroom do not always define what they mean by "process" and "higher order thinking." Moreover, as Lauren Resnick (1987) points out, thinking skills tend to resist the precise forms of definition we have come to associate with the setting of specified objectives for schooling. Still, it is relatively easy to list many key features of higher order thinking; and perhaps it is important to begin to develop "process lists" if we are to have a balanced curriculum. Resnick herself has provided a broad characterization of higher order thinking which we think is a step in this direction. It reads as follows:

- Higher order thinking is *nonalgorithmic*. That is, the path of action is not fully specified in advance.
- Higher order thinking tends to be *complex*. The total path is not "visible" (mentally speaking) from any single vantage point.
- Higher order thinking often yields *multiple solutions*, each with costs and benefits, rather than unique solutions.
- Higher order thinking involves *nuanced judgment* and interpretation.
- Higher order thinking involves the application of *multiple criteria*, which sometimes conflict with one another.
- Higher order thinking often involves *uncertainty*. Not everything that bears on the task at hand is known.
- Higher order thinking involves *self-regulation* of the thinking process. We do not recognize higher order thinking in an individual when someone else "calls the plays" at every step.
- Higher order thinking involves *imposing meaning*, finding structure in apparent disorder.
- Higher order thinking is *effortful*. There is considerable mental work involved in the kinds of elaborations and judgments required (Resnick 1987, 3).

Although this characterization does not help us decide how to teach higher order thinking, it does make clear that higher order thinking skills are more recognizable than some educators admit.

Still, a great deal of work remains to be done before we resolve the problem that Resnick poses as follows:

> Thinking skills tend to be driven out of the curriculum by ever growing demands for teaching larger and larger bodies of knowledge. The idea that knowledge must be acquired first and that its application to reasoning and problem solving can be delayed is a persistent one in educational thinking. "Hierarchies" of educational objectives, although intended to promote attention to higher order skills, paradoxically feed this belief by suggesting that knowledge acquisition is a first stage in a sequence of educational goals. The relative ease of assessing people's knowledge, as opposed to their thought processes, further feeds this tendency in educational practice.
>
> Periodically, educators resist this pressure by proposing that various forms of process- or skill-oriented teaching replace knowledge-oriented instruction. In the past, this has often led to a severe de-emphasis of basic subject matter knowledge. This, in turn, has had the effect of alienating many subject matter specialists, creating pendulum swings of educational opinion in which knowledge-oriented and process-oriented programs periodically displace each other, and delaying any serious resolution of the knowledge–process paradox . . . We need both practical experimentation in schools and more controlled instructional experimentation in laboratories to discover ways of incorporating our new understanding of the knowledge–reasoning connection into instruction (Resnick 1987, 48–49).

Higher Order Skills and Teaching Practice

This need for a deeper resolution of the content–process debate should not conceal the fact that we have learned a good deal about conveying higher order thinking to students. Here we want to take particular note of a recent report that reviewed three remarkably successful programs for teaching higher order reasoning in reading, writing, and mathematics (Collins, Brown, and Newman 1989). This review was conducted to identify common elements in teaching higher order reasoning across these three diverse disciplines. A major finding was that there are many commonalities and that many of these common elements focused on integrating content and process. Moreover, the pedagogical strategies used in the three programs attempted to convey many of the higher order reasoning skills that can be observed in the practice of experts. As stated in the introduction to the report:

> While schools have been relatively successful in organizing and conveying large bodies of conceptual and factual knowledge, standard pedagogical practices render key aspects of expertise invisible to students. In particular, too little attention is paid to the processes that experts engage in to use or acquire knowledge in carrying out complex or realistic tasks. Where processes are addressed, the emphasis is on formulaic methods for solving "textbook" problems, or on the development of low-level subskills in relative isolation. Few resources are devoted to higher-order problem solving activities that require students to actively integrate and appropriately apply subskills and conceptual knowledge . . . As a result, conceptual and problem solving knowledge acquired in school remains largely unintegrated or inert for many students (Collins et al. 1980, 454).

The report identifies several components that support the development of higher order thinking skills. We will briefly discuss two of these that have particular relevance to science instruction. The report labels these components *content* and *method.* The content component is further subdivided into four subcomponents: domain knowledge, heuristic strategies, control strategies, and learning strategies. It therefore includes what we have called *process* within its definition of *content.*

- *Domain knowledge* includes the concepts, facts, and procedures associated with a particular discipline. The report emphasizes the importance of integrating domain knowledge within realistic problem-solving situations in the domain. Examples of domain knowledge in solving algebraic word problems include definition and use of variables, algebraic rules for manipulating equations, and methods for translating English sentences into algebraic expressions and equations.
- *Heuristic strategies* are the "tricks of the trade" in a discipline and are generally acquired through practice. An example of a problem-solving strategy in algebra is to use two simultaneous equations with two unknowns to reach a solution.
- *Control strategies* consist of methods for selecting and managing problem-solving strategies effectively. Simply possessing problem-solving strategies is not sufficient, because a student may not know how to select the appropriate strategy to solve a

problem. Therefore, control strategies serve this managing function.

- The last subcomponent, *learning strategies*, refers to strategies for learning the two previous subcomponents and includes techniques for exploring the domain (for example, reading books, observing successful problem solvers), and techniques for extending and restructuring the domain knowledge as the need arises (e.g., challenging misconceptions, rebuilding knowledge, and building a hierarchical structure for the domain knowledge).

The second component, method, can be thought of as the pedagogical strategies that can be employed by teachers to give students the opportunity to observe, engage in, and discover expert strategies within appropriate contexts in a given discipline. Six subcomponents of method are: modeling, coaching, scaffolding-fading, articulation, reflection, and exploration.

- The first, *modeling*, consists of helping students build a conceptual model of the processes "experts" use to solve problems. This involves providing students with opportunities to observe the processes used by experts in solving complex problems. One possible way to do this is for the teacher to "think out loud" while solving a complex problem for the class so that all thought processes are made visible and explicit for the students. In algebraic problem solving, for example, the teacher might describe how to extract the essential features of a word problem, showing explicitly how one goes about translating them into mathematical terminology, and detailing the procedures for solving the translated problem. Equally important is extracting from the detailed procedure a qualitative summary that outlines the process by which the problem is solved.
- *Coaching* involves guiding students engaged in problem solving in such a way that they bring their performance closer to the performance of expert practitioners. Coaching seeks to integrate process and content and includes activities such as helping students identify previously unnoticed aspects of a problem that are crucial for reaching a solution. In algebraic problem solving, this might include the teacher's pointing out to a student that an important part of the word problem has been ignored in constructing a solution plan.

- *Scaffolding-fading* involves providing just enough support to allow students to carry out tasks on their own, and gradually removing the support structure as students gain proficiency. For example, the teacher may perform an intermediate step in solving a problem if it is too difficult for a student, or may provide a hint that will assist a student in continuing to solve a problem when stuck.
- *Articulation* refers to any method employed to help students verbalize their knowledge, reasoning, and problem-solving processes. Articulation not only can help the teacher monitor student progress and diagnose difficulties (for example, misconceptions) but can also help students realize that there is a close relationship between the ability to articulate and depth of understanding.
- *Reflection* refers to helping students build their own cognitive model of expertise by having them compare their own problem-solving processes with those of fellow students and those of the teacher or other experts.
- *Exploration* involves encouraging students to engage in their own problem-solving activities independent of specific assignments.

Summary Comments

To summarize the content–process debate, there is a movement to increase the emphasis given to process skills in the classroom. Content advocates believe that this movement will give rise to reductions in the content knowledge that students will receive and view this reduction as detrimental given the ever-increasing amount of scientific information available. Process advocates claim they are merely attempting to reach a proper balance in the curriculum. On the one hand, content alone cannot turn students into proficient problem solvers; on the other hand, process cannot be taught outside some body of content knowledge. Content and process enjoy a delicate symbiotic relationship—a relationship that must be nurtured.

Despite the debate, there are issues on which advocates for either side must agree. Students in this country are lagging behind those from other industrial nations in problem-solving skills. To maintain our security, financial independence, and competitive edge we must reverse this trend. What is insidious is that the ramifications of this lag are not immediate. We will face the results of our policies sometime in the future; thus there does not seem to be a sense of urgency in dealing with the problem. It is also clear that by and large our schools, by international standards, are quite competent at teaching content. It, therefore, makes sense to focus more attention on helping to bridge the process–content gap by focusing attention on ways to integrate process and content across the entire curriculum.

Finally, it is clear that integrating content and process will require close cooperation among science and mathematics teachers. Perhaps this is the biggest challenge. Classroom techniques, experiences, and ideas must be shared in order to emphasize a common approach in science teaching. An isolated science teacher emphasizing the proper content–process balance in an isolated school cannot achieve the desired global impact. Furthermore, it is important that teachers do a better job in integrating mathematics and science instruction. Mathematics is the backbone of all the sciences, yet it is taught largely in isolation. If math and science teachers collaborate to show how mathematics is constantly used in scientific endeavors, then students would not only become more motivated to learn mathematics and science but also be better able to build mental structures integrating both disciplines.

Beginning Science Education Earlier

A curriculum issue important for all science teachers, regardless of grade level, involves the importance attached to quality science instruction from the very early grades forward. For a nation whose economy depends so much on its technological and scientific expertise, the amount of time spent on science instruction in grades K through 8 is deplorable (Shymansky, Kyle, and Alport 1982; Helgeson, Stake, and Weiss 1977). State-generated objectives for the elementary curriculum mandate that large amounts of time be spent on the basic competencies: language arts and mathematics. Most would agree that these basic academic competencies are the building blocks of all knowledge and thus important to acquire early. However, there seems to be a tacit assumption that these basic competencies cannot be learned within the context of science.

Actually, science can be a powerful motivator for teaching the basic competencies. An interesting science topic can motivate students to do more reading on the topic, with the possible added benefit that they learn how to use the library. Students may also improve language skills if asked to write essays on the material they read. Finally, mathematics could be introduced as the basic tool of all science.

As an example, an elementary school science program designed for limited-English-proficiency students in which groups of students interact in a hands-on scientific-inquiry approach resulted not only in students learning science but also in measurable increases in students' English proficiency (De Avila 1988). Thus, science can serve as a "hook" to draw students' interest. Once captured, this interest can be channeled into practice in reading, writing, and mathematics.

There is a second and equally important reason for beginning science instruction as early as possible and that is to interest more students in pursuing science-related careers. Projections of the number of engineers and scientists that will be needed by the end of the century indicate a drastic shortage. Postgraduate science and engineering programs are becoming increasingly dependent on foreign nationals to fill available slots. If this situation worsens, our future technological and financial security will be dependent on a work force educated outside this country. To curb this trend, we must begin to interest elementary school students in science. Achieving this goal is largely dependent on having sound K through 6 science curriculums available, and on having a cadre of elementary school teachers who are adequately trained to teach science.

Encouraging Independent Learning

Because students are being asked to learn ever-increasing amounts of scientific information in high school and college, and then often to unlearn and relearn information due to changes later in their working lives, the major responsibility for becoming scientifically literate ultimately rests with the student. The curriculum, therefore, must reflect the aim of helping students to become self-learners. In point of fact, the earlier self-learning begins, the greater the chances of academic and work success later on. This is particularly the case where thinking skills are concerned, for, as Resnick reports:

> The most important single message of modern research on the nature of thinking is that the kinds of activities traditionally associated with thinking are not limited to advanced levels of development. Instead, these activities are an intimate part of even elementary levels of reading, mathematics, and other branches of learning—when learning is proceeding well. In fact, the term "higher order" skills is probably itself fundamentally misleading, for it suggests that another set of skills, presumably called "lower order," needs to come first. This assumption—that there is a sequence from lower level activities that do not require much independent thinking or judgment to higher level ones that do—colors much educational theory and practice. Implicitly at least, it justifies long years of drill on the "basics" before thinking and problem solving are demanded. Cognitive research on the nature of basic skills such as reading and mathematics provides a fundamental challenge to this assumption. Indeed, research suggests that failure to cultivate aspects of thinking such as those listed in our working definition of higher order skills may be the source of major learning difficulties even in elementary school (Resnick 1987, 8).

Science curriculums that nurture independent learning incorporate the following among their aims:

Encourage the Use of All Available Resources

In all science courses, students should be encouraged to draw upon a variety of resources including textbooks, libraries, teachers, other students, experts, and family members. A habit as simple as reading the textbook is not one that students adopt naturally (Mestre 1988); so it is important for teachers to encourage students to read and learn from textbooks and to go in search of library books to supplement topics in which they have an interest. Using teachers, other students, local experts, and family members as sources of information and expertise is an excellent habit to instill in students and one that will serve them well when they enter the collaborative environment of the workplace.

Teach Skills Needed to Acquire Scientific Knowledge

Students should be encouraged to ask, as well as answer, questions. Also, students should be helped to distinguish between those questions that can be meaningfully posed and answered within science and those other *nonscientific* questions that are outside the purview of science. In addition, students should be taught the art of formulating scientific questions, breaking up "large" questions into more manageable "smaller" questions, and designing a procedure (whether involving experimental work, library research, or other means) for answering them. Finally, students should be helped to make unbiased observations of the world around them, a skill that is crucial to all science and to learning in general.

Build a Self-Consistent Knowledge Base

As we suggested in the previous chapter, students commonly have contradictory views of scientific conceptions. They should be helped to perceive that these contradictory views are a natural result of ongoing learning and that it is difficult, if not impossible, to avoid such contradictions in any genuine learning process. However, students are better served if we show them more explicitly how to recognize and deal with contradictory scientific conceptions. We should instill in students a sense that science strives to be self-consistent in the way it builds knowledge. Thus, when students recognize conflicting scientific ideas, they should know the importance of analyzing the logical consequences of each in search of some flaw, thus striving to resolve the contradiction. If a resolution cannot be achieved by direct reasoning, then the student should see that, through the process of science, an experiment could be designed and performed to resolve the contradiction. Only by challenging deep-rooted beliefs will students be able to build a flexible, expandable, and self-consistent knowledge base in the scientific disciplines. Teaching students to seek consistency in building scientific knowledge in the face of apparent contradictions is perhaps the biggest challenge in helping students become self-learners.

IV. Teaching Science

This chapter provides examples of how the pedagogical ideas discussed in the previous two chapters can be implemented in the classroom. However, it is important for the reader to keep in mind that the examples are taken out of context—a context that encompasses individual teaching style, students in a class, and the curriculum within which the particular science course is embedded. These examples are intended to be illustrative and should not be interpreted as the final word. Instead, we have attempted to provide a "flavor" for teaching scientific inquiry that makes the student an active participant and the teacher a mentor-facilitator.

Kinematics

We assume that the activities described below are carried out over several days. The actual time needed to accomplish each objective will depend on the specific teacher and the specific class. More important than the actual topics and activities is their logical progression. The activities are constructed so that students will derive both quantitative and conceptual understandings of specific aspects of kinematics. The activities will also illustrate ways to probe for, and to help students to overcome misconceptions by helping them realize how some of their conceptual views contradict the findings of the experiments they perform. Thus, students are asked to deal with their own misconceptions by seeking consistency between experimental findings and their own understanding.

To accomplish this, let us look at two related activities. The first is an experiment focusing on the question: "What is the acceleration, g, of a falling body?" The experimental findings will set the stage for the second activity. This activity starts with a writing exercise and culminates in a classroom discussion, both delving into a conceptual question that has consistently been difficult for students: What is the acceleration of a ball thrown vertically up at the instant when it reaches its maximum height?

In these activities, we assume that students have covered the basic kinematic relationships, namely $d = d_0 + v_0t + (1/2)at^2$, and $v = v_0 + at$, and are able to solve textbook problems involving these equations. This topic appears early in a typical high school physics curriculum.

Activity 1: Measuring the Acceleration Due to Gravity

The activity begins with a structured inquiry approach to the nature of the earth's gravitational attraction. The initial classroom discussion focuses on encouraging students to consider and verbalize how gravity might work. Everybody has noticed that objects fall down, not up. Yet understanding the nature of gravity requires a resolution of many questions: Does gravity always pull with just one strength or does it pull more, or less, at certain times? Does gravity act equally on all objects? Do heavier objects fall faster than lighter objects?*

In all probability, teachers will find that their students have a wide diversity of views on these matters. Discussions of these questions can encourage students to propose experiments testing their own views and theories. Here, the teacher will probably have to suggest ways of refining the proposed experiments, but students should feel the experiments are theirs, not the teacher's. Groups of students may want to perform different experiments and report their findings back to the class. An alternative is to guide the class toward designing a single experiment.

To illustrate some of the activities that might lead to designing and performing an experiment using the structured inquiry approach, we will now focus on the question of whether or not objects fall with the same acceleration.

The equipment is quite simple: a stopwatch, a steel ball, and a 15-meter tape measure. Although several experimental approaches

*Recent physics research into the nature of "the fifth force" is attempting to answer this very question. It appears that the theory predicts that gravity affects protons and neutrons slightly differently, thus raising the possibility that all objects may not fall with the same acceleration. However, the effect, if present, would be excruciatingly minuscule.

may be suggested by students, we will focus on what is likely to be the most common suggestion, namely to drop the ball from a specific, known height and measure the time it takes the ball to reach the ground. If students do not mention it, the teacher should point out that if the ball falls at a constant speed, it will obey the kinematic equation, $d = d_0 + v_0 t$, but that if it accelerates, it will obey the equation $d = d_0 + v_0 t + (1/2)gt^2$. The latter equation assumes that the value of the gravitational acceleration near the surface of the earth, g, is a constant—an assumption students may not want to make.[†] Their experiment should reveal whether or not this is a valid assumption. The next step is to guide students into devising a method for systematically measuring height-time pairs over as wide a range of heights as possible. Students may need to use the school's stairwells to achieve the desired range.

Another important point that should be considered during the initial classroom discussion is how students will analyze their data. Assuming that they have rejected the possibility that the ball maintains the same speed throughout its fall, students will now need to consider how to use their data and the equation, $d = d_0 + v_0 t + (1/2)gt^2$, to extract the value of g. First, what is an appropriate coordinate system? Students will be inclined to pick the origin of the coordinate system at the point where the ball is released. This is not a wise choice since the origin will change for each height chosen. Thus, picking the origin of the coordinate system at the ground is the wiser choice. The initial height, d_0, would then always be the height from which the ball is dropped. Furthermore, (t) is measured until the ball hits the ground so the height, d, associated with this time is always 0. Finally, because the ball is dropped from rest, the initial velocity, v_0, is also 0. These substitutions simplify the kinematic equation describing motion under a constant acceleration to: $0 = \mathrm{d}_0 + (1/2)gt^2$.

After the teacher helps the students derive this equation, the

[†]The assumption that g is constant holds for the falling ball as long as 1) air resistance does not play an important role, 2) the ball remains near the earth's surface, and 3) the earth has a uniform density in the vicinity of the experiment. The first two conditions are satisfied with this experiment, and although the last condition may not hold, its effect is extremely small.

teacher can cover a subtle but important point. Because this equation states that the sum of two quantities is 0, then the fact that the first quantity, d_0, is always positive in the chosen coordinate system means that the quantity, $(1/2)gt^2$, must be negative. Because the square of a real number, t^2, cannot be negative, then g must be negative. This is an important conclusion because a negative value of g implies that the direction of the gravitational acceleration is down toward the earth, not up toward the sky. This subtle point is bound to cause confusion, and teachers should be prepared to discuss it further. To avoid possible confusion, the teacher may choose to write the equation $0 = d_0 + (1/2)gt^2$ as $d_0 = (1/2)|g|t^2$. In this form, all quantities in the equation are positive. Indeed, it is the absolute value of the gravitational acceleration, $|g|$, that we wish to extract from the data.

The final topic to consider during the initial classroom discussion is how the data will be analyzed to extract a value for $|g|$. Students are most likely to suggest using $d_0 = (1/2)|g|t^2$ to obtain a value for $|g|$ for each height-time measurement, and then averaging all the individual measurements of $|g|$ to obtain a "best value." This method will yield an average value for $|g|$ but will *not* give an indication of the value's error range. A simple method that gives a good estimate of both the "best value" of $|g|$ and its error range involves a graphical analysis of the data. To use this method, students should make a table of their data with entries for four different quantities: d_0, t, t^2, and $(1/2)t^2$. If the acceleration due to gravity is both constant and independent of height (assuming no air resistance and that we stay relatively close to the earth's surface), plotting d_0 on the y-axis and $(1/2)t^2$ on the x-axis should yield a straight line with a slope that gives a value for $|g|$ based on the measurements. Thus, students can plot their data and draw the "best line" starting at the origin and going through the data. They can also obtain a reasonable estimate of the error in the data by drawing two additional lines, one steeper than the "best line" and one less steep, such that these two lines enclose about two-thirds of the data points. The slope of the steepest line will provide the highest value of $|g|$ consistent with the data, while the slope of the least steep line will provide the lowest value of $|g|$. Figure 3 illustrates this procedure. Selecting the graph's axes skillfully in order to obtain a straight line is a very powerful ana-

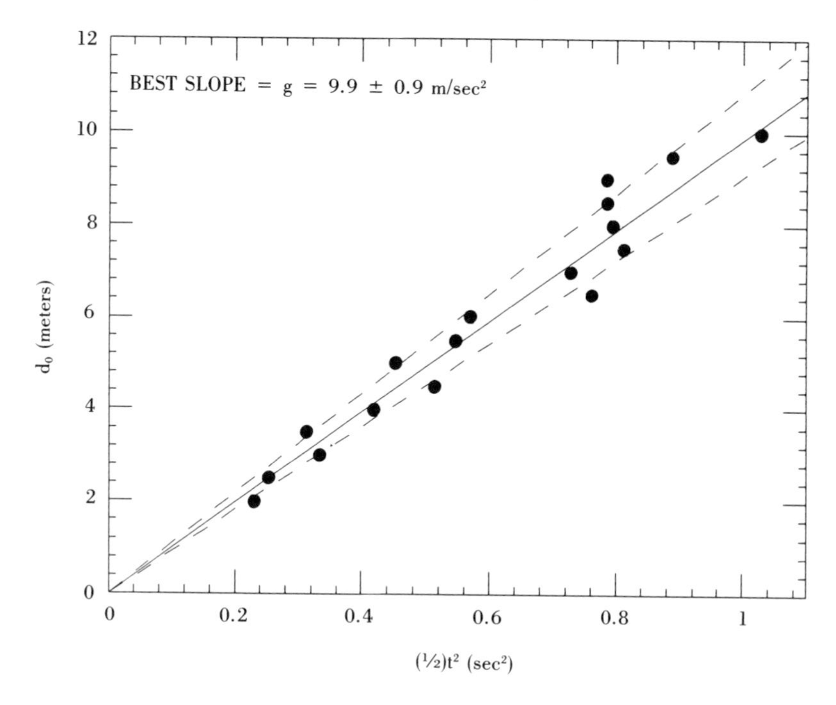

Figure 3. Graphical procedure for obtaining a "best fit," upper and lower limits to the gravitational acceleration data.

lytical technique. The technique may appear "magical" to students and cause varying degrees of confusion. However, if the technique is repeated several times throughout the school year, students should eventually come to understand the reasoning behind it.

Clearly, the initial discussion will take more than one class period. Indeed, we could not begin to write down all the possible questions that could arise. Carrying out the experiment and analyzing the data will take another class period. The structured inquiry discussion following the experiments and the data analysis can be used to highlight issues related to designing approaches that minimize error, to determine how well the measured values agree with the "accepted" value of 9.8 m/s^2 (or 32 ft/s^2), and to highlight other related experimental methods or theoretical issues. For example, students might be asked what the results might have been if the earth's gravitational acceleration was not constant, but varied with height.

Activity 2: Probing and Dealing with a Common Misconception

The laboratory activity described in the previous section provides an excellent background for studying a common misconception concerning the acceleration of a vertically thrown object when it is at the top of its trajectory. Many students associate zero velocity with zero acceleration. Because the velocity at the top of a vertical trajectory is zero for an instant, students often believe that the acceleration must also be zero at that instant. After students complete the previous experiment, the teacher can probe effectively for this misconception with the following assignment:

Write a paragraph describing the speed and acceleration of a steel ball that ascends vertically to some maximum height and then drops down to the earth again. Make sure you discuss the speed and acceleration of the ball at different points in its trajectory, especially at the point where the ball reaches its maximum height.

On the day this assignment is due, the teacher can have several students read their paragraphs. The ensuing classroom discussion can be used to examine and deal with the "zero velocity implies zero acceleration" belief.

After several students have read their paragraphs, the teacher can begin to make a table on the blackboard summarizing the students' views. After all views are adequately described, the teacher should encourage students to debate opposing views within an orderly classroom discussion. Clearly, the role of the teacher during these discussions is to monitor the classroom dialogue and offer some guidance at appropriate times to ensure progress. As mentioned in Chapter II, an effective means for students to deal with their misconceptions is to point out disparities between their beliefs and observed physical phenomena. The teacher can do this in a number of ways, such as by asking questions, for example: "If the acceleration suddenly becomes zero at the top of the trajectory when the ball's velocity is also zero, then what does the kinematic equation, $v = v_0 + at$, imply about the subsequent velocity of the ball?" Or, "If the ball is experiencing a nonzero acceleration just prior to reaching its maximum height as well as just after it starts to come back down from its maximum height, then how did the earth know to 'turn off gravity' for just an instant?" Students can also be challenged to seek the simplest explanation for the observed phenomena. At various times during the classroom discussion, the teacher can take straw votes to evaluate whether or not students with erroneous beliefs show signs of confronting their misconceptions. These votes force students to commit themselves to a specific viewpoint, regardless of whether or not it is correct, and to evaluate that viewpoint.

Students' belief that "zero velocity implies zero acceleration" may stem from an inability to understand the definition or meaning of acceleration. The concept of acceleration is not as intuitive as that of velocity. For example, we can observe velocity simply by gauging how fast something is moving. However, it is much harder to judge whether or not an object is accelerating by observing its motion. To compound the confusion, we can "feel" acceleration when a vehicle in which we are riding suddenly changes speed, but we cannot "feel" velocity when the vehicle is moving in a straight line at a constant speed.

Several thought-provoking arguments should help students consider the meaning of acceleration. For example, one argument draws on the experiment in Activity 1: "The instant the ball is released, it has zero velocity; if it also had zero acceleration at that instant, then it would not move since the kinematics equation governing its motion states that $d = (1/2)at^2$." Another argument can draw directly upon the definition of acceleration as a measure of the change in velocity with time: "Consider the case of the ball that went straight up and down. If the ball had a tiny upward velocity just prior to coming to a stop at the top of the trajectory, and a tiny downward velocity just after coming to a stop at the top of the trajectory, then the velocity was clearly changing during that time interval; therefore, if the velocity changed, there had to be acceleration at the top of the trajectory because acceleration is defined as the change in velocity over the time interval in question."

We offer these arguments not so that students can be "hit over the head" with them, but as springboards for discussion. The teacher can move the discussion in the direction of one of these arguments by using leading questions, such as those given above, or by asking the class to explore the logical consequences of a particular statement made by one of the students. There should be no expectation that once one of these arguments has been articulated every student will immediately accept it. On the contrary, research has shown that considerable discussion over a period of days or weeks is often necessary before students can begin to develop confidence in these ideas (Minstrell 1982, 1987).

Respiration

There is a temptation to talk about the latest, most newsworthy findings in science. These are not always the most interesting topics for students and it is usually very difficult for the teacher to handle them in a way that allows students to explore and experiment. In the previous section we saw how the most basic of topics, namely motion, can provide an environment for honest investigation. Nothing is more fundamental to our own existence as biological entities than breathing. Simple respiration can give students an exciting window on the essence of biology. We will now consider a range of investigations that can be made using the simple apparatus shown in Figure 4.

The Warburg respirometer works on the principle that soda lime (CaO) will absorb carbon dioxide by removing it from the air above it ($CO_2 + CaO \rightarrow CaCO_3$). If an animal is placed in the respirometer it will take in oxygen and exhale CO_2. Because the CO_2 is absorbed, there will be a net decrease in the volume of the air equal to the amount of oxygen consumed by the animal during respiration. This decrease can be measured quite accurately through the following process. As the volume of the air decreases, the level of the water in the monitoring manometer will fall. The original level can be restored by ejecting air from a syringe into the respirometer. The amount of air needed can be read off a scale on the syringe. A typical first experiment involves placing a mouse or frog in the respirator and measuring the consumption of oxygen

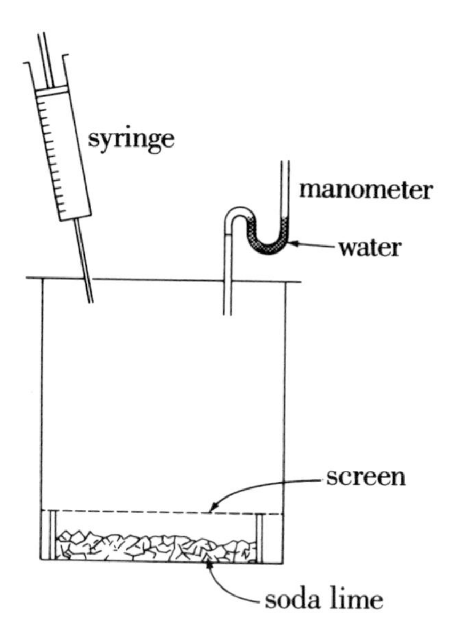

Figure 4. Warburg respirometer.

over a stated period. (Readings may be taken every minute.) In most cases, students will need a good deal of guidance in combining the volume readings and the time readings to develop a measure of the respiration rate ($\Delta V/\Delta t$). Rate is a difficult concept for many students and is often not taught carefully before calculus—if then. Therefore, it is a good idea for the teacher to demonstrate several different ways of carrying out this calculation. The volume readings on the syringe can be plotted on a graph of volume versus time; the slope of the graph gives the rate. On the other hand, individual changes in volume can be divided by the time interval (usually one minute) over which they occurred. This method will produce many different readings and help to emphasize the sources of error. Students will notice that time readings are not exactly one minute apart and that the change in volume readings, ΔV, involves two uncertainties. Finally, when the rate concept has been thoroughly mastered by the students, the teacher can explain the even more abstract notion of respiratory rate per unit mass. Mathematically, the calculation involves no more than dividing the respiratory rate by an animal's mass. Measuring the mass of a moving mouse presents certain problems in experimental technique, but the harder issue is conceptual. What is a volume per unit of time per unit of mass? Because changes in rates are common in all science, it is prudent for the teacher to spend some time discussing this issue to make sure the students understand what they have measured. Such a discussion is likely to consider how rate changes may vary for different kinds of animals and this, in turn, may suggest a whole range of new experiments.

In the respiration experiment, there is a rich body of material that could be used within a structured inquiry format to allow students to make interesting measurements. By taking some additional time, a potentially cookbook experiment can be turned into a serious laboratory experience that addresses many of the outcomes that we described in Chapter II. For example, a meaningful, answerable question (outcome A.1) might be "how do we know that the CaO absorbs CO_2 or that it absorbs all the CO_2?" Students should be allowed to pose such questions and then think of ways to experimentally answer each question on their own. They might, for instance, suggest filling a container with CO_2 and then determine what happens when they add CaO (A.3 and A.4). The role of working hypotheses (A.2) may become clear as students recognize that eventually they must take on faith that the source of CO_2 they use (whether a compressed gas or a chemically generated one) does, in fact, give them CO_2.

In exploring this sort of question, the teacher can raise the issue of controls. Might the volume of air in the respirator have decreased without the mouse? If so, we need a control jar that has no mouse. Perhaps the mouse raised the temperature in the jar and breathing had nothing to do with the effects we observed. As a control, we could place a hot rock in the apparatus. (The issue of how large and how hot a rock to add could raise many interesting questions. New concepts that can be introduced through the rock experiment include heat capacity and rates of heat radiation.) The issue of controls and appropriate observations is one route to consideration of the points listed in outcome B, gathering scientific information.

Drawing valid conclusions (outcome D) may be illustrated by considering the issue of whether or not oxygen is the element that the breathing mouse removes from the jar. We could compare and contrast respiration to placing a candle in a bell jar (Figure 5). After the candle has gone out, a considerable amount of water comes into the jar, but contrary to what some textbooks say, this is largely the result of cooling the air rather than replacing the missing oxygen. The candle goes out long before all the oxygen is used up. We can demonstrate this hypothesis by adding wet steel wool to the jar and leaving the jar to stand overnight. By the next day, the steel wool will have rusted (a process requiring oxygen) and additional water will have entered the jar.

There are, of course, many other questions that students may want to ask about the original experiment. For example, some might want to place a plant in the respirator to see whether it "breathes." Here, it is important to distinguish between a plant in daylight carrying out photosynthesis and a plant in darkness. In the first case, the plant takes in CO_2 and gives off O_2; in the second case, the opposite process occurs.

Before students begin the experiments, the teacher should determine the general time frame for reaching interesting results. Students need guidance or they may never achieve a successful experimental outcome. This is an example of the type of help that distinguishes structured inquiry from a pure discovery situation.

By now, what had started out as a simple hour-long experiment has raised many interesting questions, most of which can be

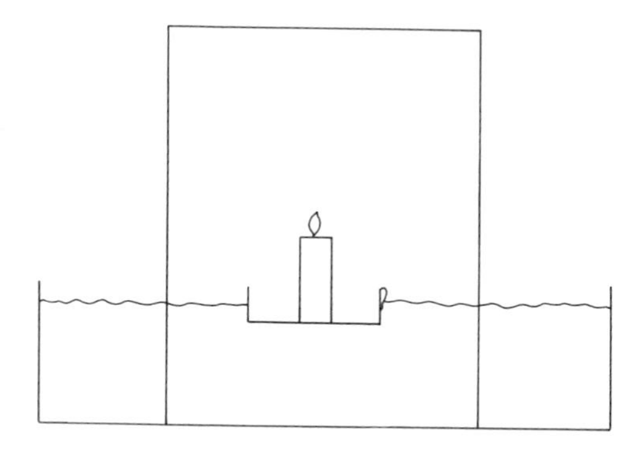

Figure 5. Burning candle inside a jar surrounded by water.

answered by the students themselves using familiar equipment. The class could be divided into different groups, each of which would research one issue and report back to the others. In some cases, groups would share information because their topics overlap. In addition, students are likely to bring up some issues that teachers would not think of. Some of the topics that could be considered include the effect of temperature, the kind of animal or plant, and the activity of the animal or plant. By separating the class into small groups it is easy and natural to emphasize the need to organize and communicate one's results (outcome C). Students should make both written and oral reports. Whether or not other students understand what is being said can be used as a standard of evaluation for grades.

Class discussion of the various experiments should automatically indicate the need to design controls and to make explicit certain assumptions about the processes being examined. These discussions will probably reveal factors that were not considered initially. These factors and the challenges of accounting for experimental uncertainties could be used to illustrate the tentative nature of scientific conclusions and the interdependence of scientific knowledge. Students can also be guided to realize that a specific scientific experiment usually raises questions that can only be answered by designing and conducting many more experiments. There is no better way to explore the role of experimental work in the construction of theories (outcome E). While discussions of this sort are likely to be heated and may even seem too enjoyable to permit coverage of serious scientific content, they are the closest we can ever come to showing students what "real" science is and to training students to think scientifically.

None of the experiments we have discussed thus far are necessarily the ones teachers should select for their own classrooms. A wide range of projects could have served just as well. What is important is that students are challenged to think about the evidence they have collected in the laboratory and are given the time to repeat and reexamine their work in the light of new issues raised by this reflection. This cannot be done when a large number of different cookbook experiments are crammed into one semester. But there are dangers in our approach. If a great deal of time is devoted to the investigation of one topic, students may become lazy or lose interest. Holding students' interest has always been a mixture of magic and experience. There seems to be no certain way to predict what will prove engaging. As a teacher gains experience with the kinds of questions students are likely to raise and discovers the experiments that can be used to answer them, it becomes progressively easier to keep labs exciting.

Chemical Properties and the Periodic Table

Activities in the previous two examples are readily recognized as "doing science." This third example is rather unorthodox and does not fit neatly into that mold. We, however, consider the activities and skills involved in this exercise just as important for integrating reasoning in science with reasoning across disciplines. We have chosen to use an example from chemistry to illustrate this approach somewhat arbitrarily and not because chemistry lends itself to such treatment better than other subject areas. At the close of this section, we suggest topics from other science disciplines that could be used with this approach just as readily.

In Chapter II we mentioned briefly the important role played by qualitative reasoning in problem solving and in the development of scientific theories. We will draw out some of the ideas introduced in Chapter II with an example of the use of qualitative reasoning. The example given here involves the historical development of the periodic law of the chemical elements and the periodic table.

There are several advantages to teaching qualitative reasoning within a historical context in science. First, it shows students that scientific advances and breakthroughs are most often the result of work that continuously builds upon a body of knowledge. If we neglect these contextual aspects in teaching science, students may think that scientific breakthroughs are always the result of an inexorable chain of careful, quantitative experimentation or theorizing by an individual working in isolation. In actuality, breakthroughs seldom occur in a vacuum—no pun intended. Scientific advances occur when conditions are "ripe." If some crucial ingredient is lacking, the breakthrough may not take place. Three additional advantages of approaching science from a historical perspective are that:

- We can integrate some of the basic academic competencies, such as reading, writing, reasoning, speaking, and listening, in teaching science.
- We can apply many of the science learning outcomes within a qualitative context.
- We can integrate science and the humanities.

In the following exercise, students are supposed to analyze the conditions and circumstances leading to the development of the periodic law of the elements. Students will have to pose and defend hypotheses about why conditions were "ripe" for Mendeleev to make his bold proposition that the properties of elements change in a periodic pattern as atomic weight (or number) increases. This task will require students to begin by exploring the historical developments leading to the periodic law. There are various general questions that will help students conduct their historical analysis. These questions include: (1) What important scientific questions were asked in epochs prior to a particular scientific advance, and why were these questions important? (2) For a given historical period what were the accepted scientific beliefs and theories? (3) Did the accepted beliefs and theories facilitate or obstruct a particular scientific breakthrough? (4) Who were the *savants*, the scientific experts, in different epochs; what did they espouse; and how did they affect the scientific breakthrough under consideration? (5) What underlying scientific, social, and political forces influenced a specific scientific field in different epochs? Which of these forces facilitated and which obstructed progress? (6) Given our current state of scientific knowledge, can we identify the ingredients that were necessary for the breakthrough or advance?

These general questions can guide us in answering the focal question: When do advances or breakthroughs in science occur? The answer to this question is crucial not only in analyzing a scientific advance that has already taken place but also in analyzing current bottlenecks in scientific progress. This latter task is extremely difficult, since we do not have the benefit of using historical fact to support our conjectures. Nonetheless, we still might be able to identify a political, social, or scientific force creating a current bottleneck.

Reasoning Qualitatively from a Historical Perspective

This activity can be used either as an introduction to the study of the properties of the elements or as a review after the subject has been covered. The activity can support both a short-term, or a long-term assignment. It is best to introduce the activity as a qualitative reasoning exercise; a clear distinction should be made between quantitative and qualitative reasoning. The arguments students will be asked to construct will not involve numbers or be subject to precise mathematical verification. However, their qualitative arguments must withstand rigorous tests of both validity and reasonability. It is important for students to appreciate that their arguments must stand on their own merits and are subject to refutation if not properly defended. The teacher can help students appreciate that the goal of the assignment is to use qualitative reasoning, in this case reasoning within a historical context, to formulate a coherent mechanism that explains how Mendeleev arrived at the periodic law. Students should be aware of some of the questions listed above, which they could use in the assignment.

One way to get students started is to assign readings about the historical circumstances leading to development of the periodic law. Students would all be armed with the same information that they could then use in their arguments, so that all students would be working from a common body of knowledge. This strategy would also facilitate the teacher's role as moderator, with the approach working best for a short-term assignment.

A second approach, best suited for a long-term assignment, consists of giving students general guidelines for what is expected of them, assigning them to teams, and asking the teams to research the periodic law and use this research to complete the assignment. This approach is preferable because it promotes the spirit of inquiry. Each team member could research a central question, and then all team members would work together to consolidate the information and to formulate the hypotheses and arguments they wish to propose. This offers an excellent opportunity for lively discussions that promote qualitative analyses and informal reasoning. To make the discussions most efficient the teacher should be present during some of each team's discussions to offer guidance and to ask pointed questions.

After team members are satisfied with their research and "answers" to the guiding questions (mentioned on page 76), they should write a joint paper proposing a coherent view of the events leading to the discovery. This should not be a "research report" that simply states the facts, but should propose a mechanism by which an advance or breakthrough takes place in science, and should back up the argument with specific examples (from their research) of the events leading to the development of the periodic law and periodic table. Each team could then be given 15 minutes to present its proposal and argue for its acceptance in front of the class. The arguments presented by the different teams will likely evoke lively discussions and counter arguments from all the students. Here, it is important for the teacher to monitor the validity of the arguments used. Because each team has spent considerable time on the project, the members will probably defend their team's theory with emotional as well as logical arguments. This gives an opportunity to contrast an argument based on fact and developed logically with an argument built on emotional appeals or personal attacks. The teacher should grade the written assignments on the depth and quality of the reasoning.

What Kinds of Historical Events Might Students Use in Answering the Guiding Questions?

In conducting the research, team members may be uncertain as to how to create the historical data base from which they will later argue their case. The teacher can provide guidance, particularly in suggesting how to go about answering some of the guiding questions listed earlier. For example, the history of chemistry could be traced to the sixth century B.C. when Greek philosophers asked a very important scientific question, namely, "Of what is matter composed?" Three philosophers—Thales, Anaximenes, and Heraclitus—proposed that all matter was composed of different forms of the same substance: Thales claimed it was water, Anaximenes air, and Heraclitus fire. In the fifth century B.C., these ideas were expanded by two philosophers: Anaxagoras proposed that all matter is composed of tiny particles, not unlike atoms as we conceptualize them today. Empedocles added earth to the previous three "elements" and further claimed that all matter was not composed of a single substance, but rather of a combination of four basic constituents—earth, air, fire, and water. The latter theory was favored for many centuries. Not until after the art of alchemy became popular did science philosophers and savants begin to gain insights into the chemical properties of matter.

The precursor to chemistry, alchemy is loosely defined as the process by which a common substance is transmuted into a substance of greater value. Two forces motivated alchemists, one was the desire to turn common metals (such as lead) to gold. The second was the obsession to find an elixir, a potion to prolong life. Because of the prevailing view that earth, air, fire, and water comprised all matter, alchemists believed that turning lead into gold simply involved figuring out the percentages of these four "elements" in lead and gold and increasing or decreasing the proportions to effect the change. Clearly, students could take very different points of view in arguing whether the four-element theory and the forces driving alchemy facilitated or obstructed progress. What is clear is that by the mid-sixteenth century, chemistry had made only slight advance.

A major shift in scientific thought during the seventeenth century can be attributed to the work of two savants who were not alchemists. Copernicus challenged the geocentric view of astronomy, in which all celestial bodies were said to revolve around the earth, and showed that a heliocentric model, in which all bodies in the solar system revolved around the sun, was better able to explain astronomical observations. No longer could savants involved in the study of natural philosophy (a name for science in those days) impose their own egocentric view on the universe. Now, the universe, or more precisely our observations of the universe, had to dictate the scientific views we took, not vice versa. This inquiry-based approach to science was also adopted fully by Galileo, who became the first true experimental scientist.

This shift in scientific thought paved the way for Boyle to provide the first meaningful operational definition of the term *element*. In order to be self-consistent, he reasoned, the definition had to describe a substance that could not be broken down into other substances by chemical means. Furthermore, compounds formed by these elements should be separable into the original substances. This view not only made better sense than the earth-air-fire-water view but it also helped debunk it. It was clear that by Boyle's definition, earth was not an element since it contained many substances. Fire was not even a substance at all! Instead, fire was a process by which substances change chemical states (although the chemistry of combustion was not known until later). In Boyle's day, nine substances fit his definition of element (carbon, copper, gold, iron, lead, mercury, silver, sulfur, and tin). Although it was Boyle who coined the term "chemistry," largely to distinguish the new methods from those of alchemy, this pseudo-science—alchemy—continued to be practiced for some time.

Students will very likely uncover one theory that they could easily argue obstructed chemistry's progress from Boyle's time through most of the eighteenth century: the *phlogiston* theory was used to explain the mechanism now known as combustion. Phlogiston was thought to be released when substances burned. According to this theory, when a substance burned it released all of its phlogiston, thereby becoming the same substance in "unphlogisticated" form. Thus, when a substance, such as a log of wood, burned, it left behind its unphlogisticated form—ashes. The unphlogisticated state was called "calx." The same mechanism was used to explain the process of rusting. Chemists then believed that rusting consisted of a metal burning and leaving behind its unphlogisticated calx, namely rust.

Another important historical figure, Lavoisier, now considered the "father of chemistry," changed the direction of scientific thinking and will probably be singled out by students. It was Lavoisier who combined various philosophical and scientific views available at the time to make significant advances. Through his efforts chemistry became a largely quantitative science. He was instrumental in debunking the phlogiston theory and replacing it with the oxidation theory we know today. He was even able to use his oxidation theory to argue that water was a compound, and not an element as the Greek philosophers had thought. Yet another important contribution was Lavoisier's suggested nomenclature for compounds that is still used today.

When the nineteenth century began, the search for elements was still a "hit and miss" game. There was no procedure for discovering individual elements. A major breakthrough occurred in the middle of that century—the invention of the spectroscope by Bunsen and Kirchhoff. The spectroscope displayed a substance's frequency spectrum (i.e., the light given off) while it burned. This decomposition of frequencies produced unique color bands (frequencies of the electromagnetic spectrum) that were the "fingerprints" of specific elements. Now chemists could search for elements by burning substances and looking at the spectral fingerprints they generated; generating a unique spectrum indicated the presence of a unique element.

A combination of the spectroscope and hard work enabled chemists to identify approximately 60 elements by the middle of the nineteenth century. However, the pace of discovery necessitated a scheme for categorizing these elements. Clearly, some schemes (e.g., arranging elements in alphabetical order) do not describe the elements' most important chemical properties, namely their ability to form compounds.

Working at the beginning of the nineteenth century, Dobereiner noticed that three elements, calcium, barium, and strontium, exhibited very similar properties. For example, they had similar melting points and formed similar compounds with other elements. A few years later, Dobereiner found two other "triads" which again had similar properties—sulfur, selenium, and tellurium; chlorine, bromine, and iodine. Although Dobereiner's observations were on target, nearly a half century would pass before Mendeleev used them to deduce the periodic law.

Because the discovery of the periodic law is treated in such a cursory manner in chemistry textbooks, students may have formed the impression that Mendeleev sat down in his laboratory and magically came up with the law totally on his own. This is far from the case. As is usual with major scientific breakthroughs, precursors had laid the groundwork. In addition to Dobereiner, Beguyer de Chancourtois published a table with elements arranged in a spiral column, with triads of three elements forming a vertical line. Another chemist, Newlands, arranged elements into columns and found larger families of elements that exhibited properties similar to those identified by Dobereiner. Thus, various chemists observed the periodicity of the elements, although it took Mendeleev to state the periodic law in unequivocal terms, namely that the properties of the elements exhibited a pattern when arranged in order of increasing atomic weights.

Mendeleev did not focus on atomic weight, or atomic density, as many had before him. He focused on the valences of elements, which measure an element's ability to form compounds. He arranged elements in order of increasing atomic weights and noticed that the valences formed a periodic pattern. In fact, Mendeleev proposed correcting certain elements' atomic weights simply because they did not fall nicely into his periodic arrangement.

At the time, many chemists were unconvinced of the validity of Mendeleev's scheme. However, one aspect of Mendeleev's table stood out. While many chemists had arranged elements in tables, all slots were filled with known elements. Mendeleev's table had "holes," because no known element fit the periodic pattern. Thus, what made the periodic law and table truly remarkable was its predictive ability. Mendeleev actually predicted the existence of several elements. That alone sounded as though Mendeleev was practicing magic rather than chemistry. However, it was in predicting the properties of these unknown elements that Mendeleev achieved total credibility. For example, he predicted the existence of the element now known as gallium. He predicted its atomic weight, specific gravity, and melting point, as well as how it would react and combine to form compounds. Gallium was discovered five years after Mendeleev predicted its existence, and its chemical properties were exactly what Mendeleev had predicted. In discussing this development, it is crucial for students to realize that a physical law or theory not only explains known phenomena but also has predictive value.[‡]

Topics from Other Sciences

The kind of qualitative reasoning exercise illustrated can be applied to many "day-to-day" topics covered in science classes; however, this approach is best suited to topics in which major breakthroughs or advancements occurred and in which there is a rich body of historical material related to the advancement. Of course,

[‡]The historical summary provided here is based on Isaac Asimov's excellent book, *The Search for the Elements* (1962).

in implementing the activity, the depth of coverage and the amount of time can vary to meet curricular constraints.

As noted, possible topics are plentiful in all areas of science. For example, in the biological sciences, the events leading to the discovery of the helical structure of DNA by Crick and Watson (Watson 1968; Judson 1979) or the events leading to the birth of immunology through Jenner's discovery of vaccination at the end of the eighteenth century are rich topics for the activities described. In earth science, plate tectonics has most likely been the biggest major breakthrough in the last twenty-five years (Uyeda 1978; Hallam 1973; Glen 1982; Seyfert and Sirkin 1979) and has met with remarkable success in explaining a host of geological phenomena, from glacier formation to earthquakes. Physics offers a host of topics for this kind of activity, including the development and consequences of Einstein's theory of relativity, the theory and experimental verification of fundamental particles (Fundamental Particles and Interactions Chart Committee 1988), and countless "classic experiments" which have served to advance theory throughout the centuries (Shamos 1987). Astronomy and cosmology offer topics such as the origin of the universe and of time (Trefil 1983; Hawking 1988) and the theory that led astronomers to explain why the sky is dark at night[§] (Harrison 1987).

The approach can also be used to analyze the social and scientific ramifications of certain landmark events, such as the Soviets' launch of the Sputnik satellite into orbit. The scientific advances in the United States spurred by that event were numerous, far reaching, and included the ushering in of our space program, the total revision of the science curriculums in our schools, and practical products such as Teflon-coated cookware, quartz crystal watches, and freeze-dried foods. Another landmark event that can be qualitatively analyzed from a historical perspective is the Manhattan Project, which resulted in the making of the atomic bomb and the birth of the atomic age.

[§]Why the sky is dark at night is not as obvious as it might seem at first blush. Since there are countless stars in the sky, and since stars have a finite extent, any line of sight we choose in the night sky should eventually intersect the surface of a star; hence we should see light in all directions. Harrison (1987) explains why this is not the case.

Regardless of the topic chosen, the benefits of the approach are clear. It promotes several useful skills, such as writing, reasoning, debating, hypothesizing, library research, self-learning, and learning from peers. In short, the approach promotes skills useful for general reasoning and learning across the curriculum.

V. Science and the Basic Academic Competencies

This chapter offers some suggestions about how the study of science should also contribute to the development of the Basic Academic Competencies identified in *Academic Preparation for College* (College Board 1983), that is, reading, writing, speaking and listening, observing, using mathematics, reasoning, studying, and using computers. It is now generally recognized that the basic competencies are most effectively learned within disciplinary contexts (Resnick 1987). Therefore, it is important to consider how the development of these competencies can be made integral and necessary to work in science. The key to this connection is reasoning, a competency that few would see as separate from science. Recent research has shown that the development of reasoning is intimately tied to skill in communication and representation (Nickerson 1982). To a great extent, reasoning is a form of internal communication in which the alternatives are clearly represented and carefully considered. Thus, reading, writing, speaking and listening are not just adjuncts to a course in science but the means by which students develop the ability to think scientifically.

Communication and Representation

Representation is the general process of putting ideas in a communicable form. The importance of communication and representation in the development of scientific reasoning is only now beginning to be understood. In older theories of learning they played an incidental role; in the constructivist view taken here, they are the essential processes through which knowledge is built. To begin with they are the essence of speaking and listening, writing, and reading. But they play a no less important role in observing, studying, reasoning, and in mathematics and computer applications. Vast new opportunities are opening up as computer-based technologies create ever more powerful communciation and representational tools (Janvier 1987). This chapter illustrates some of the ways in which science teachers can help their students gain scientific competence through greater communication and representational facility.

Speaking and Listening

Speaking, listening, and thinking are the basic elements of effective communication. Classroom discussions in which students play a very active role are keys to the development of students' scientific reasoning. Science teachers have properly attached great value to precise explanation. But if students are to develop an active understanding of science (see the discussion in Chapter II), they must have the opportunity to formulate and express their ideas. Initially, their expressions will be groping and tentative, not sharply drawn. Teachers should encourage students to explore their ideas and come to terms with their preconceptions. Teachers also need to avail themselves of opportunities to listen. Such attentive listening involves interpretation and attempts to identify the assumptions implied by what students are saying about natural phenomena.

Listening is not necessarily a natural activity for teachers. The wait-time studies (Rowe 1973) show that we often give students less than one second to respond to our questions. The simple act of waiting a few seconds before teachers answer their own questions can in itself heighten the level of student reasoning. Still greater progress is possible if we listen to several student responses before providing our own views or the right answer. To do this we must control body and facial language, as well as our tongues.

An accumulating body of evidence also shows that many teachers ask the most able students to speak in class more often than other students. This may keep the hypothetical ball rolling and ensure that what is said in class is close to what we hope for, but it does not provide students with the opportunities they need to formulate and express their emerging understanding of the natural world. As students are encouraged to formulate and express their own understanding of the natural world, that understanding is likely to become more precise and coherent. In fact, some of the most effective science teaching (Minstrell 1987) depends on this drift toward coherence. In classes of this kind the teacher's role is not to guide students toward some predetermined conclusion but rather to make sure that each student's ideas are fully considered by all the class (for example, the teacher might ask, "but what about Molly's suggestion? How does that fit with what Paul said?") Though time-consuming, extensive discussions are often the only way to reach understanding of a topic with deeply rooted misconceptions.

Small group discussions in which the teacher plays a mainly managerial role are often used to promote thinking skills. Experience gained from such teaching shows that if the questions are appropriately framed, a student discussion will make reasonable progress with relatively little guidance from the teacher (Minstrell 1987). Such discussions are likely to consider many alternatives that would not have occurred to someone experienced in the field. Indeed, some student preconceptions may seem dangerously off base. However, only by helping students think through their naive, previously unchallenged ideas can teachers promote a solid basis for understanding.

There are several ways in which both teachers and students can develop the speaking and listening skills needed to carry out such classroom discussions. To some extent, teachers can simply model the questioning and listening that they want. But to attain the highest levels of skill, programs such as reciprocal teaching (Palincsar and Brown 1984) and pair problem solving (Whimbey and Lochhead 1987) are useful. These programs provide specific exercises that help students learn the appropriate skills component by component. Teachers who want to become highly competent facilitators of student discussions may choose to gain experience conducting and analyzing videotaped clinical interviews. This can help them develop the skill of "getting inside the student's head." A second valuable training exercise involves repeated viewings of a videotaped class discussion to understand the complexity of the multilevel arguments students use.

Writing and Representation

In the science classroom, writing generally has been limited to preparing laboratory and field reports and to answering examination questions. But a major change has been occurring, and some science teachers have altered both how writing is approached and how it is used in teaching science. They approach writing as "thinking on paper" and report that it improves students' retention of what they have learned. This approach begins with the insight that writing is not just a matter of applying some accumulation of "mechanics" to preexisting thought. Rather, writing is seen as a process of formulating, clarifying, and refining both thought and expression.

Science teachers should find this approach quite useful. Science students already do a fair amount of ad hoc writing, recording laboratory and field observations and taking notes on readings, lectures, and discussions. A teacher can take advantage of this by asking students to jot down their thoughts on some question raised during a lecture (Strauss and Fulwiler 1987). These jottings can give information about students' thinking or can merely be something for the student to either keep or throw away. Seeing writing as a process helps to fit these activities into a larger whole (Hayes and Flower 1986). An expert on writing might stress that in getting words on paper the student has already begun the process of writing, a flow of activity that leads to fuller understanding and expression. The process approach to writing can substitute a streamlined flow for the notorious writer's block. From the viewpoint of the science teacher, the larger whole into which these student writing activities fit is, of course, the overall process of scientific thought. Chapter II of this book tries to clarify the role of observation and description in that regard.

Both the process of scientific thought and the process of writing involve organization and integration. Each aspect (thought and writing) can be mutually refining. Sometimes student laboratory or field reports can be accurate with regard to particulars but may not logically lead to the stated conclusions. Answers to examination questions can be rich in information but never really illuminate a specific point. Learning to approach writing as a process could help students make the transition from information to conclusion. For example, a student who has learned how to invest some effort in getting observations and notes organized on paper could come to class with an examination answer already "prewritten" to a significant extent. The writing done on the examination would be a recollection, an extension and a refinement of that preparation.

We are describing a "process" approach to writing that stresses revision. Assignments are not written once, but are revised to clarify and sharpen each thought. As they review drafts, teachers may identify commonsense assumptions that interfere with students' grasp of scientific concepts and explanations. The process of revision may help students recognize and come to terms with these obstacles. Indeed, writing as a process can be so useful in developing a coherent understanding of science that some teachers are beginning to assign long-term projects that involve a great deal of intermediate writing and eventually lead to a more refined product than is usually achieved in science classes.

Many science teachers, of course, do not feel prepared to help their students approach writing as a systematic process and may want to collaborate with members of the English faculty. Such an intercurricular strategy could begin with students writing papers describing everything they did in preparing their most recent written work in the science class: examination, laboratory report, or long-term project. By working with the science teacher, the English teacher would be able to analyze the students' accounts and identify those activities that were most effective as preparation for carrying out the written work in science. Feedback from this analysis would help students strengthen their approach to writing in science and also help identify the strategies, particularly the prewriting activities, that lend themselves to writing and thinking in science. If such an assessment were conducted on a regular basis, students' ability to "think on paper" would improve. Competence in writing about the subject will also improve students' understanding of the science.

Of course, writing is but one form of scientific representation. There are numerous other forms, including mathematics and computer-generated systems. Learning to express ideas in various representational systems is increasingly important in science. These representational media are not always easy to use, but cognitive research is revealing ways to improve student understanding. For example, a study using Newtonian free-body diagrams (Heller and Reif 1984) showed that specific instructions telling students how to enumerate all the forces that act on a body can significantly raise student performance. Other studies show how representations that we tend to think of as straightforward require substantial instructional attention. For example, research on students' understanding of elementary algebra indicates that equations as simple as $y = 6x$ are often misunderstood (Lochhead and Mestre 1988). The confusing aspect of algebraic representations can be seen by noting that the equivalence statement 3 ft = 1 yd is represented algebraically by $3y = f$ where y is the number of yards and f the number of feet.

An interesting computer-generated representational system called Thinkertools has proved very successful for teaching the basics of Newtonian mechanics to sixth-grade students in a way that enabled many of them to obtain a higher level of understanding than that normally attained by high school students (White and Horwitz 1987). A key part of the program required students to pay attention to the relationships between different representations. Figure 6 shows part of a Thinkertools exercise. The ball, which repre-

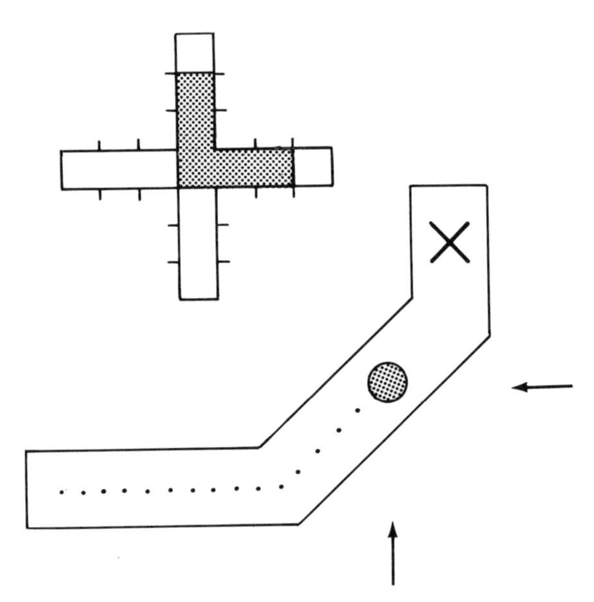

Figure 6. Thinkertools display.

sents a "rocket," is moving diagonally along a "docking ramp." The student's job is to "dock the rocket" at the bay marked *X* without crashing the rocket on the walls of the docking ramp. The components of the velocity of the rocket are represented by the shaded portions in the "data-cross" on the upper left; in this case, the rocket's velocity is made up of two equal units along the $+x$ and $+y$ directions, which give the resultant diagonal velocity describing the ball's movement. Other representations include the actual motion of the rocket along the docking ramp; the spacing of the dots in its wake, which represent the position of the ball at specific, equally spaced time intervals; and the sound of the joystick-generated "rocket bursts," which provide acceleration bursts to the rocket in any of the four Cartesian directions. (Each rocket burst results in one unit of velocity being imparted to the rocket in the specified direction.) The accompanying instructional materials include exercises that explicitly ask the students to use each representation and compare it with the others. Some exercises, for example, require that students navigate an off-screen rocket back to the screen; the only guides are the "data-cross" to indicate the rocket's current velocity and arrows that point to the position of the rocket. (When the rocket is off screen, the arrows are "pegged" against the appropriate edge of the screen.)

A computer representational language called Stella (Mandinach and Thorpe 1987) and several ecological and population simulation programs provide representational tools that enable students to think about areas of science that, until recently, could only be analyzed by those who had mastered differential equations. Such computer-aided representations are playing increasingly important roles in both mathematics and science. It is now as important to introduce students to these tools as it is to prepare them in mathematics and in laboratory practices.

Qualitative graphing is another important area in representation. This involves sketching the rough shapes of graphs, reading such shapes, and relating the graphs to a series of events. Cognitive research has shown that students rarely develop this ability merely by plotting numerical graphs (Clement 1985). Meaningful graphical interpretation is dependent on qualitative analysis.

One exercise that forms a good introduction to qualitative graphical analysis is the following: Sketch the shape of a speed versus horizontal position graph for a bicycle trip in which you start off on level ground and then go up and down two long hills (see Figure 7). Students often draw a speed versus horizontal position graph that reflects the shape of the two long hills, when in fact, as Figure 7 illustrates, the actual speed versus horizontal position graph has the opposite features. The key element in these exercises is to require students to move among various different representations, including graphs, data tables, and English sentences. Figure 8 illustrates an exercise to help them hone these skills.

Reading

If class time is devoted to discussion and laboratory or fieldwork, there is little time left for a standard lecture. As a result, a great deal of knowledge must come from reading (Wolf 1988). This is as it ought to be, but realistically there are a few obstacles. The first is that teachers must find some way to require that the reading actually is done. One approach involves structuring the class so that students cannot function without reading. A direct approach

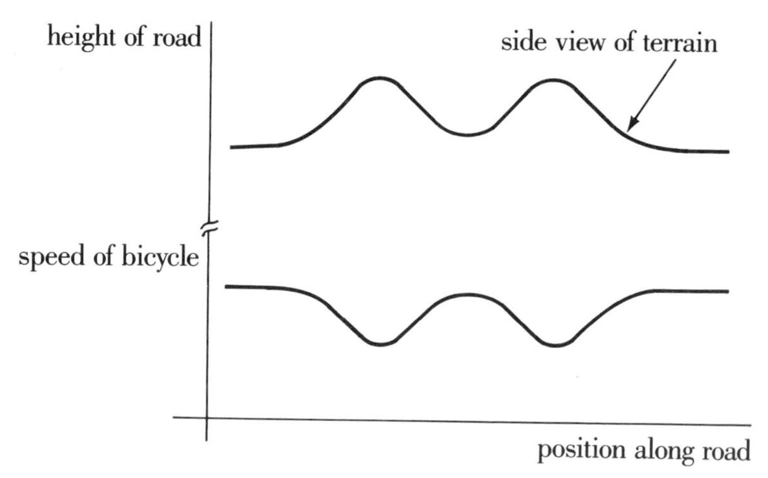

Figure 7. Qualitative speed versus position graph from "terrain view" of road for a bicycle trip.

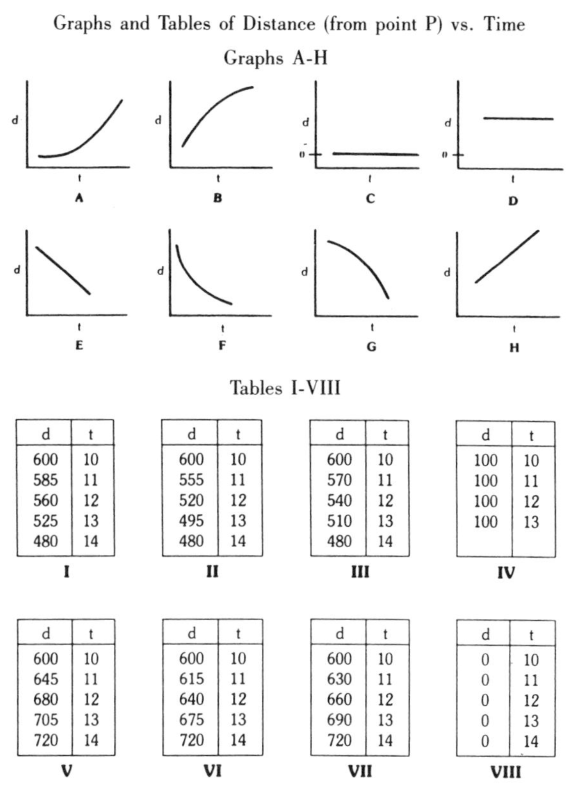

I

d	t
600	10
585	11
560	12
525	13
480	14

II

d	t
600	10
555	11
520	12
495	13
480	14

III

d	t
600	10
570	11
540	12
510	13
480	14

IV

d	t
100	10
100	11
100	12
100	13

V

d	t
600	10
645	11
680	12
705	13
720	14

VI

d	t
600	10
615	11
640	12
675	13
720	14

VII

d	t
600	10
630	11
660	12
690	13
720	14

VIII

d	t
0	10
0	11
0	12
0	13
0	14

The graphs and tables above show distance from point P as a function of time. The units for both distance and time are arbitrary. For each question, indicate the one graph and one table that best show the motion described.

Figure 8. Exercise to help students understand graphical representation.

involves giving quizzes on the readings. A second and more difficult obstacle to overcome concerns textbooks, many of which are not suitable for all students. Even if the book is written at a level appropriate to the grade being taught, it may be neither interesting nor comprehensible to students. Finding a textbook that is both readable and interesting can be a major undertaking.

The task is only somewhat easier if students can be taught to be better readers. Improving reading skills is worthwhile because those skills have so much to do with scientific reasoning. As researchers in this area have said: "Students must be encouraged to process actively information presented in the text, make predictions, draw inferences, and evaluate the quality of written material" (Raphael 1987). One technique that can help to develop these skills is a vocabulary map, as depicted in Figure 9 (Raphael 1987). Students are asked specific questions about a new term: What is it? What is it like? What are some examples? This process is similar to concept mapping (Gowin and Novak 1984). Here students face a more open-ended task and the job is to relate many different words in a semantic network. Depending on context, certain connections between the terms will be defined by the teacher (for example, "in," "of," "example of," "requires"). Students will then gradually add to their map as the course content increases. Another increasingly popular approach is to ask students to write down and hand in all the reading that they did not understand.

All of the above techniques are best used to produce material that can be profitably analyzed by the students themselves in small groups. These sessions help to promote various good habits, such as increased student interest in reading, thinking, and reasoning, and an increased ability to construct the student's own understanding of science.

Observing

Probably no aspect of science is more important or more difficult to teach than the skill of careful observation. At one time, scientists thought it was possible to make an objective observation, in the sense of seeing things in the one right way, even though it might take some practice to become good at it. We now recognize that observation is inherently theory dependent (Kuhn 1970). Each of us observes things differently, but those who observe with the "spectacles" of a common theory do see similar things.

Because we now recognize that many students come to us with theories that are quite different from our own, we cannot expect

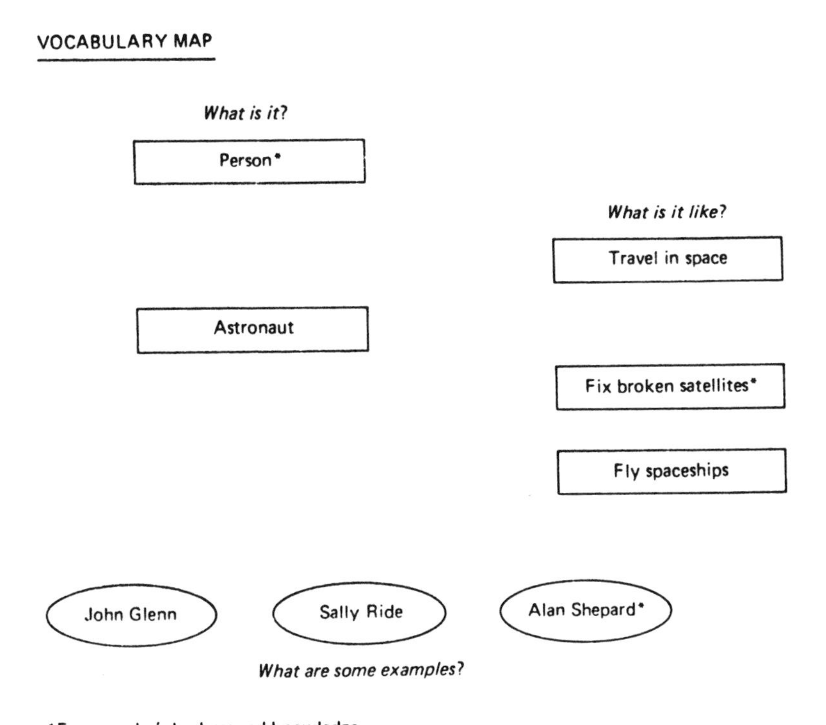

Figure 9. Examples of vocabulary mapping.

them to make the same observations that we do. Often students must change their theories before they can learn to observe scientific phenomena differently. As a result, teachers are caught in a double bind.

If, as is often the case, we grade students' observational reports based on textbook notions of what they *should* have seen, we may be asking them for a report that does not match what they believe they saw. By rejecting the students' perceptions, we place them in a situation where it is safer for them to "observe" what is in the textbook than to trust their own observations. Therefore, to teach scientific observation teachers must accept each student's unique observations. Attention to detail and precision of description may be criticized, but the description of what was seen must be accepted. We are in the position of a news reporter interviewing a reputable eyewitness who has seen a flying saucer land. We can ask for lots of detail or push for evidence, but we must accept the observer's belief in his or her own observation. Eventually, however, we must move beyond the chaos of what the untrained eye sees. One method for doing this is to hold postlab discussions. What makes an observation scientific is that it is reproducible, by the initial observer and by others. Science depends on a process through which the members of a community of observers learn to make observations about which they can all agree. For students, coming to a common agreement about their observations is critical but unfortunately, often neglected.

Many will argue that the degree of student independence that we are calling for in the above paragraph is a recipe for disaster. How is it possible for an untrained group to reach the desired consensus? This is indeed an important question. Teachers must select topics for observation in which the likelihood of eventual agreement is high. They must also monitor student discussions to point out possible contradictions, to suggest new perspectives, and sometimes, to provide gentle guidance.

Consider, for example, the situation in which students are asked to observe the motion of a cart that is catapulted along the top of a table by means of a stretched rubber band (see Figure 10). Students may be told to record their observations on a qualitative graph of speed versus time and to indicate what forces are acting on the cart. The interesting point is that while neither of the speed versus time graphs shown in Figure 10 is correct, a discussion between adherents of the two views is likely to create a graph that *is* correct. Although students may be poor at criticizing their own position, they will find weaknesses in the perspective of other students. Once both positions have been undermined, the group is free to find a solution they can agree on, one free of the weaknesses of the original views. If this new view is not entirely correct, it is nonetheless likely to be closer to the correct view than either starting position was.

Another approach to inducing self-correcting observational strategies is to have individual students observe separate parts of a process and then have the group as a whole describe the com-

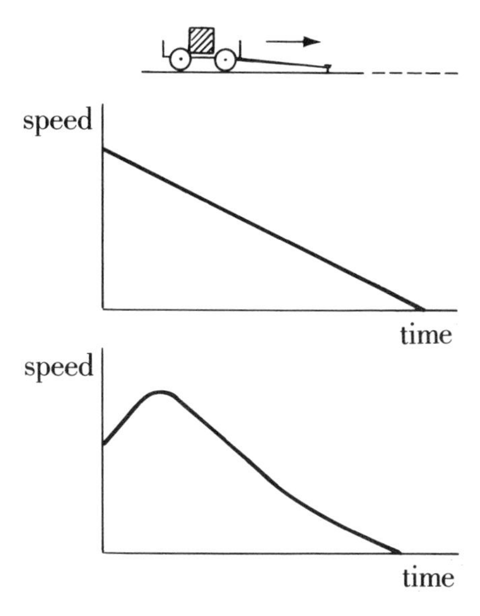

Figure 10. Two common incorrect speed versus time graphs drawn by students depicting motion of a cart catapulted by a rubber band.

plete event. A race around the outside of the school building might be observed by four or more separately placed individuals who would each sketch a qualitative position versus time graph of what they saw. This graph would show the relative positions of the runners, including when one runner overtakes another. The graphs could then be combined to form one large graph of the entire race. At this point there is likely to be some argument because the separate graphs will probably not agree.

Other self-correcting situations could include the following. The behavior of a laboratory animal might be observed by different students over different time spans and a complete picture pieced together. Here a count might be made of the number of food pellets eaten which could be compared to the number missing from the food bin. Alternatively a plant or chemical that is responsive to light might be described by different observers at different times and then an attempt made to integrate the observations. This would likely lead to the need for further and more careful observations.

Mathematics

Mathematics is a part of science in the broadest sense, and science is partly mathematics. The old boundaries of deductive versus inductive investigation are beginning to blur as computer simulation becomes a more and more important tool in both fields. Today it is possible to find biologists and mathematicians using the same methods. Now, more than ever, there is a need to coordinate the teaching of mathematics and science.

Physics and mathematics have always been close. To a great extent, the needs of the physics curriculum have determined what was taught in mathematics. But as more fields demand math, and as new forms of mathematics enter the curriculum, the needs of the physics curriculum exert less influence. Physics teachers may have to teach more math; they will also have to pay attention to changes in the math curriculum and learn to take advantage of new techniques, modifying their curriculum to exploit new opportunities. Fractals and chaos theory are but two of these new areas in which new mathematical techniques abound.

Chemistry and biology have not been as close to mathematics, but that situation is changing rapidly. The need for close cooperation among chemistry, biology, and mathematics teachers will be great as new topics with which everybody is unfamiliar enter the curriculum.

Reasoning

In a sense, this entire chapter is about the use of science to teach reasoning, so there is no need for a separate section on the subject. But there is one aspect of the issue that deserves special attention, the conflict that can arise between teaching scientific facts and scientific reasoning. This happens when students use a valid premise to reach unacceptable conclusions.

As an example of the fact versus reasoning dilemma consider how one should respond to the student who says that there are more daylight hours in the summer than in winter because heat makes things expand? Apart from the obvious overgeneralization of a principle, the student has applied the principle unequally; were it a valid idea, the principle should apply equally to daylight and darkness. The teacher might ask this student whether there are more hours of darkness in summer as well; the student might answer "yes"—from his or her perspective it may well seem that way. In this case, a teacher cannot really fault the student's reasoning because no one can be expected to reach correct conclusions without having all the facts. Unfortunately, in subjects as complex as modern science, we can never expect students to know all the facts. On the other hand, our student may recognize that the nights are shorter in the summer and see a need to revise the explanation. We might be disappointed that the student failed to notice this objection without help. But can we really expect students to take everything into account? Certainly this is not a reasonable expectation before students have had plenty of experience with a particular topic. Finally, what about confusing cause and effect? Shouldn't the student have recognized that the cause and effect factors were backwards? Again, the answer is, "not necessarily." Piaget found that young children often say the wind is made by the swinging motion of trees and, given their knowledge, this is not an unreasonable hypothesis; it at least indicates an attempt to integrate their observations.

In the above example we have deliberately picked a humorous and unlikely case for emphasis, and to provide perspective. Realistic examples require that we step back from what we know to be true and try to see things from the student's perspective. Consider, for example, the following proposed experiment. To test whether the unit of calorie used in food analysis is the same as that used in physics and chemistry, a flask containing 100 cubic centimeters of oil with a known caloric content is heated from 30° C to 40° C. A similar flask of water is heated in the same way on the same heat source and the heating times are compared. Because we know that we need one calorie to raise 1 cubic centimeter of water by 1 degree centigrade, we can calculate that the water has absorbed 1,000 calories. If the oil takes twice as long to heat up, it must contain 2,000 calories. We can compare this figure with the number given in a table of food caloric contents and see whether they agree. This experiment is reasonable in every respect but one, namely, it fails to recognize that the calories in food have to do not with heat capacity but rather with the energy released through digestion. It is a critical mistake that shows a profound lack of understanding, and yet the experiment shows some sound scientific reasoning that ought to be applauded. In point of fact, this experiment was proposed not by a high school student, but by a Ph.D. in psychology. If this were a student report, how should we have graded it?

There is an apocryphal story about a scientist who wanted to show that grasshoppers hear with their legs. He trained a grasshopper to hop every time he rang a bell. Then he cut off the grasshopper's legs and it no longer hopped when he rang the bell. The conclusion is true, even though the experiment is seriously flawed: grasshoppers do hear with their legs.

Just as we cannot take a student's incorrect answers as evidence of poor reasoning, we also cannot take correct answers as evidence of good reasoning. We must look at the entire process and understand the student's assumptions. This is difficult and time-consuming, but often very interesting.

Recent research shows that reasoning from incorrect hypotheses is a natural and essentially unavoidable part of learning science. Thus, in teaching scientific reasoning, we must find ways to encourage correct thinking even when the initial assumptions are in error. This can only be done if we know what the student's assumptions are. It has always been a good practice to insist that students state their assumptions and describe all the steps in their reasoning. Now that we know more about the details of the learning process, it is even more important to make such demands.

Studying

Reasoning requires discussion, which usually takes place in groups. Homework study groups can help to extend reasoning beyond the classroom, and teachers must play an active role in forming and maintaining such groups. First of all, teachers should encourage formation of study groups by indicating which assignments should be done in groups and which should not. Next, it is often useful to set the task of forming a group as an assignment in itself. Rules need to be established covering group size (say three to five members), leadership, meeting times, and so forth. Groups need to learn how to enforce discipline on themselves; for example, every member must come to each meeting fully prepared. During meetings, everyone must actively participate. One highly effective style is to have a student solve a problem by thinking aloud while other members of the group question and challenge each step. Such sessions need to be modeled for the students; it is often helpful if the teacher sits in on the first few meetings.

A useful technique is telling members that each of them is responsible for how all team members learn. Quiz grades might be based on the average performance of the group, or one team member may be called on at random to present a problem solution to the entire class. (To avoid negative reactions from students and parents, it is important to make group-graded exercises a small part of the total grade.) It is also important to stress that few jobs these days require work performed in isolation and that cooperative problem solving and learning is commonplace.

Study groups can also encourage reading. Students may be told to review the readings in their groups to prepare for a quiz. If they know the teacher will not be covering the material in class prior to the quiz, the chance that they may actually read the assignment is better. Each member of the group might be responsible for paraphrasing a part of the reading, with the rest of the team passing judgment on its validity, completeness, and emphasis.

Groups are also a powerful tool for encouraging writing and editing. Members of a team might read and correct each other's papers before submitting them to the teacher. If several short papers have been written, for example daily assignments, the group might help each individual decide which papers to turn in for grading. In either case, the teacher's grading load is reduced and the students get valuable experience.

Using Computers

Computers and calculators are becoming increasingly important in all areas of society. Initial concern over the possible adverse ramifications of using computers and calculators is giving way to the realization that we had better decide how to use them effectively in the classroom. National commission reports call for the use of calculators and computers at every level of the mathematics curriculum. Undoubtedly, they will soon be required in many testing situations (National Research Council 1989).

Calculators are a mixed blessing: they can confuse as well as simplify. Consider, for example, the topic of exponential growth. A teacher could ask students to use the exponent key to produce a series of numbers representing a population's growth over time. In some cases, this approach might work well, but many students do not know about exponents, and others scarcely understand what they do know. For such students, using the calculator only makes the topic even more mysterious. Alternatively, repeated multiplication by 2 will demonstrate the doubling of a population over time. Here, the calculator clarifies by representing a complex calculation as the repetition of a simple one. Once this pattern is well understood less cumbersome techniques can be introduced to those students who are ready for them. Teachers should not underestimate the value of having their students repeatedly and repetitiously calculate a novel process the long way. Calculators are of tremendous value here, because doing such calculations manually requires a dedication bordering on obsession.

Computer simulations provide the same sorts of opportunities and dangers as using a calculator. But they can allow consideration of certain topics that would not be feasible otherwise. Computer simulations can establish an environment where students explore and gradually develop a sense of how a system works. On the other hand, simulations may hide essential features in a system and make mysterious what should be transparent. Ideally, users should see computers as tools to help them accomplish some objective. Computers are much less effective when used for demonstrations. The point here, as Seymour Papert has indicated, is that we should be asking what kids can do with computers rather than the reverse (Papert 1980).

Computers are becoming an integral part of standard laboratory equipment. But once again, the educational benefits do not follow automatically. The electronic pan balance that digitally displays the numerical value of the mass resting on it does nothing to help students understand the concept of mass or even how it is measured. This situation is in striking contrast with an electronic thermometer connected to a computer in such a way that the computer displays, in real time, a graph of temperature versus time (Malone, Nalman, and Tinker 1984). Admittedly, while this thermometer is also of little help in teaching how temperature is measured, it does provide a powerful tool for observing how temperature changes with time (and, because the thermometer can be moved, how temperature changes with position in space). It is quite simply impossible to generate real-time graphs in any other manner. Without an opportunity to observe a physical situation in such a direct way, students are likely to lose the sense of the dynamic process that a time graph supposedly represents. As a general rule, automation hides some processes but, if used cleverly, brings others to light. It is important for teachers to be aware of how technology is affecting what their students are seeing and to wisely choose how to use it. Such considerations often suggest new areas for investigation such as asking: "How does an electronic scale measure mass?" or "How could we make the mass measurement without such a scale?"

VI. Toward Further Discussion

The illustrations and observations we have presented are meant as suggestions for improving the quality of science instruction. Because our suggestions are quite general, they must be adapted to a particular classroom setting and to a particular teacher's style. This adaptation will require further discussion among science teachers. Moreover, there are many additional issues that science teachers, and those who work with them, must consider; and these will be discussed in this chapter. Although we have not been hesitant to state our own opinions on these issues, we are well aware that many science teachers may hold other points of view. Our purpose in this chapter is not so much to persuade as to provide the starting point for an ongoing dialogue among colleagues.

Attracting, Preparing, and Retaining a Competent Cadre of Science Teachers

The preservation of democracy depends on an educated populace able to make well-informed decisions. Moreover, both our national security and our ability to compete economically in international markets depend on a technologically sophisticated work force. Yet, proportionately, we devote few resources to education. Of particular concern to us is that so many science teachers feel that they must leave the profession if they are to garner financial rewards, respect, and independence. To attract and retain highly qualified science teachers will require not only higher salaries but also the creation of conditions in which teachers have more control over and responsibility for what they teach. Educational improvement will require long-term planning, including the gradual establishment of strong bonds between the local community and the school. This will be possible only if the teaching staff remains relatively stable. Yet, with the current shortage of well-trained science teachers, school administrators often are forced to react to short-term crises (for example, finding someone to teach high school physics by September) rather than planning for long-term success.

This shortage of science teachers has created an urgent situation in which, increasingly, teachers in one science discipline are asked to teach classes outside their immediate field of expertise. This, of course, is undesirable. Even an excellent teacher will be less than effective teaching a subject in which he or she is not adequately trained. Providing students with the appropriate perspective in a discipline is a task requiring substantial training and experience in that discipline. While recent cognitive research has defined some strategies that can be generally applied across disciplines, these do not obviate the need for subject matter expertise. Therefore, a displaced biology teacher teaching a physics course needs substantial assistance. Without such help the teacher may have no alternative but to focus strictly on textbook content, ignoring problem solving and reasoning. It is far easier, of course, to convey factual knowledge that can be found in books than it is to convey problem-solving techniques that require the appropriate background and practice.

There is no solution to this problem short of attracting an adequate pool of qualified teachers, but it would help if we could develop imaginative approaches to in-service education to address this problem. Resources for doing this might be created through collaboration among school systems, universities, and government. Community and business involvement are also desirable, particularly in districts that serve large numbers of disadvantaged students. It should be possible for several small school systems to form a consortium to deal collaboratively with teacher-training and enhancement. Such a pooling of resources would increase the chances that the financial backing and expertise needed are available. If a particular school system needed to retrain teachers outside their areas of expertise, the consortium could provide the necessary training by a master teacher from one of the consortium schools. Forming partnerships with universities is another way to help retrain teachers outside their disciplines. Yet a third avenue is to have teachers apply for enhancement support from government agencies such as the National Science Foundation.

Obstacles to Science Learning at the Elementary School Level

We have previously mentioned the importance of beginning science instruction in the early grades. Indeed, many excellent K-6 science curriculums have been in existence since the 1960s, including: Science-A Process Approach (SAPA) developed by the American Association for the Advancement of Science, Science Curriculum Improvement Study (SCIS) developed at the Lawrence Hall of Science, and Elementary Science Study (ESS) developed by Education Development Center. These programs all involve teaching science via process-oriented, hands-on laboratory activities. The curriculums are accompanied by kits containing necessary "equipment." Although studies indicate that science teaching can be more effective using these curriculums than it is with traditional textbook approaches (Helgeson, Stake, and Weiss 1977), the curriculums have not been widely adapted and have been plagued with problems in many everyday classroom settings.

One problem is that the hands-on approach these curriculums represent requires that teachers be specially trained in that approach—training that the majority of elementary school teachers never received. In addition, using a hands-on approach involves more time and expense than traditional textbook approaches. Because they contain expendable supplies, kits need to be maintained, requiring money and time, both luxuries that compete with curriculum areas currently having higher priorities than science, such as language arts and mathematics. In practice, kits were often bought to be shared among several teachers in order to save money. Scenarios whereby a teacher would spend time setting up an experiment only to find that supplies necessary for the experiment had been exhausted were not uncommon. The frustration arising from these circumstances resulted in science kits finding their way to storage closets where they were soon forgotten.

The alternative to using a hands-on approach to teaching elementary school science is the more traditional textbook approach. Although it is not optimal for illustrating the experimental nature of science, the textbook approach can be used effectively to teach several important aspects of science. For example, posing meaningful, answerable questions, making accurate observations, and solving problems can be achieved within a textbook approach. The key problem is the temptation to teach science simply as a body of facts. Needless to say, the soundness of the textbook is an important determining factor in the success of such an approach.

Although either approach can be used, it is clear that the teacher's ability to convey the authentic flavor of science depends largely on how well that teacher is prepared to teach specific science subjects. Unfortunately many elementary school teachers' college experiences in science have left them intimidated, frustrated, and ill-prepared. But there are ways of handling the teacher education dilemma at the elementary level. One is to provide elementary school teachers with a well-rounded college curriculum in both the biological and the physical sciences, but with an emphasis on the same type of active involvement that we hope to see in elementary school science instruction. If this were to happen, teachers would not only be better prepared to teach science as a process of inquiry but they also would be better able to integrate science with language arts and mathematics so that students see links among reading, writing, mathematics, and science.

Another approach is to train science specialists (either via pre-service or in-service programs) who would work with the classroom teacher to teach science. This model is not without precedent in elementary schools since students already receive instruction in music, art, and physical education by subject-matter specialists. The advantages of using science specialists include having a teacher adequately trained in the subject matter and having a guarantee that science is taught in the elementary grades on a regular basis. Possible disadvantages of this approach include: 1) the danger of creating the impression in students' minds that science is not something everyone can do, since it is taught by a "specialist"; 2) the additional difficulty of coordinating efforts with the classroom teacher to integrate science activities with the rest of the curriculum; and 3) the necessity to restrict science coverage to 45-minute periods several times a week, thereby excluding the possibility of performing experiments or undertaking science projects that take longer than the alloted time. Nevertheless, given the weakness of science instruction at the elementary grades, we believe that the advantages of moving to science specialists outweigh the possible disadvantages.

Redressing Imbalances

The number of female and minority college students majoring in science has increased in the past decade, but these groups still remain significantly underrepresented both in the study of science and in the pursuit of scientific careers. This is a troubling fact both for reasons of equity and for our national interest. We must take note that demographic projections indicate the number of American youth is shrinking dramatically. Between 1980 and 1996, our youth population, ages 15–24, is expected to fall 21 percent, from approximately 43 to 34 million. At the same time, minority youth are becoming a higher proportion of this dwindling population. Therefore, it follows that if we cannot interest many more minority students in science, and increase the number of females as well, we will fall short of being able to fill the nation's scientific and technical positions from our own citizenry (Wetzel 1987). As a report of the National Science Board states:

> The Nation is not being adequately served by current efforts to increase the number of women and minorities in the science and engineering work force. Unless these efforts are maintained where they are effective and intensified where they are not, the nation will continue to deprive itself of an important source of future scientists and engineers to offset the decline in total number of new entrants expected between now and 1995. (NSB 1986, p. 19)

Since minority students drop out of school in disproportionately high numbers and take comparatively fewer mathematics and science courses, we need to tackle the problem of underrepresentation early in these students' educational careers. We suggest that future attempts to ameliorate the situation focus on the following goals:

- Change the prevalent perceptions among many women and minorities that science and mathematics are alien subjects, exclusive and forbidding, and that an ability to learn mathematics and science is totally innate and does not depend on hard work and effort.

- Instill in all students the notion that a career in science, mathematics, or engineering is rewarding and within their reach.
- Improve the science and mathematics offerings in schools that serve large minority populations.
- Encourage parents and role models to support minority children in pursuing science and mathematics education.
- Encourage the corporate sector, as well as community-based organizations, to initiate and support programs designed to interest and encourage minority students in technical and scientific professions.

Effectiveness of Textbooks in Science Instruction

The textbook is an important factor in science instruction. Unfortunately, most science textbooks appear to be cast from the same mold. Why this is so is not at all clear. We are unaware of any compelling research indicating that the style and format currently used are optimal for learning. Nonetheless, a typical science textbook presents material, provides a few worked-out examples illustrating the application of the concepts and procedures covered, and lists problems at the end of each chapter for student practice. This style appears to have emerged simply to mirror traditional instructional approaches. Some research indicates, however, that a deep conceptual understanding of science may be better developed if textbooks are modified so that they place more emphasis on showing students how to solve problems through worked-out examples (Sweller 1988; Sweller and Cooper 1985). Thus, textbooks might be improved if they relied more heavily on worked-out examples to illustrate the material and concepts covered. At the very least, this research suggests that it is very important to take a fresh look at how textbooks are designed and developed.

Assessment of Learning Outcomes

Thus far, we have not discussed the assessment of student achievement in science at any length. This is not meant to downplay the importance of assessment, for we think it deserves a great deal of attention from teachers. Indeed, a full treatment of issues in assessment might easily double the length of this book. We do, however, want to say a word about tests of science achievement currently available to teachers and schools. It is important, of course, that teachers understand for themselves which outcomes are assessed in these tests as well as which ones are not. Tests designed to measure mastery of content will provide entirely different information from tests designed to measure conceptual understanding and problem-solving ability. Therefore, before we use or act upon the results of tests, we need to know if they assess what we deem important.

This simple commonsense rule presents some difficulties for the point of view expressed in this book—namely that conceptual understanding and problem-solving skills should be given much more emphasis in curriculum and instruction. The fact is that most tests in science now on the market do not measure these outcomes adequately. These tests are predominantly of the multiple-choice kind, and although some higher order skills can be measured through multiple-choice questions, the strength of such tests is in assessing factual knowledge rather than conceptual understanding or procedural skill. To the extent, then, that current tests drive curriculum and instruction, we face a serious obstacle to making needed changes in science education.

The good news is that it is possible to make better tests, and teachers can help this to come about sooner rather than later by insisting that tests really measure what students should know and be able to do. We need to stress again and again, for example, that it is more important for students to be able to pose a hypothesis that explains the findings of an experiment than it is simply to provide a statement of what the findings were. Likewise, it is more important that a student be able to perceive the principle involved in solving a physics problem, and to devise a procedure for applying that principle, than it is that he or she be able to plug numbers into formulas to produce a "right" answer. Once we as teachers make distinctions such as these clear, we will have taken an important step in promoting the development of more appropriate assessment instruments.

Moreover, we also can improve the climate for change by emphasizing that assessment should be integrated much more closely with classroom instruction. Too often, we think of assessment as something that happens only at the end of instruction, rather than as a means of assisting work with students as it is developing. Recent studies (Mestre et al. 1989; College Board-ETS 1989; Lipson 1988; Glaser 1988; Hardiman, Dufresne, and Mestre 1989) show, however, that assessment techniques can be devised which provide teachers with much more immediate information about how a student is thinking, thus making it possible to adjust instruction to suit individual needs. This, of course, is much preferable to leaving evaluation to the end when it is often too late to help a student correct misunderstandings or overcome weaknesses.

The Need for a Continuing Dialogue

Teachers can become isolated in their own classrooms and laboratories; yet the vitality of a profession depends on a continuing dialogue among colleagues. *Academic Preparation for College* (College Board 1983) was created through such a dialogue, and the intent of this volume is to expand this discussion to include many more teachers. Organizations of science teachers at the national and state levels provide for such dialogue, but these can seem remote from the daily work situation of a high school science teacher. Nevertheless, if a continuing dialogue is to exist, it has to have its roots in the cadre of teachers that comprise the individual science departments in high schools across the country. There are many themes on which such a dialogue might be based, some of which we have tried to develop in this book.

One area receiving additional attention and which, we believe, holds much promise is active, in-class research on learning. In most cases, such research will require collaboration between science teachers and a local university. There is an enormous amount that we do not know about how students think about the topics we cover in science and the effectiveness of different approaches to developing student understanding. The only possible way to thoroughly investigate these issues is through the participation of large numbers of classroom teachers. School systems must recognize the need for such studies and reduce teaching loads for those teachers actively conducting in-class research. It is also essential that the federal government support such research activity, particularly in those schools where resources are stretched thin. It is in such schools that we find students who need help the most, but about whom we know the least.

Creating ongoing discussion and collaboration among science teachers is particularly important in the context of this book. This entire book is intended to stimulate a fresh dialogue focusing on a new conception of science teaching and learning. It will have no impact unless science teachers discuss the new ideas and try them out in the classroom in the best spirit of problem solving.

Bibliography

Adler, Mortimer J. 1982. *The Paideia Proposal: An Educational Manifesto*. New York: Macmillan Publishing Company.

American Association for the Advancement of Science. 1989. *Science for All Americans*. Project 2061. Washington, D.C.: AAAS.

Asimov, I. 1962. *The Search for the Elements*. New York, N.Y.: Basic Books.

Beasly, W. 1985. "Improving Student Laboratory Performance: How Much Practice Makes Perfect?" *Science Education* 69: 567–576.

Boyer, Ernest L. 1983. *High School*. New York: Harper & Row.

Boyer, Ernest L. 1984. As quoted in Thomas Toch, "For School Reform's Top Salesmen, It's Been Some Year." *Education Week* June 6.

Brown, S.C. 1958. "Do College Students Benefit from High School Laboratory Courses?" *American Journal of Physics* 26: 334–337.

Carey, S. 1986. "Cognitive Science and Science Education." *American Psychologist* 41: 1123–1130.

Cavin, C.S., E.D. Cavin, and J.J. Lagowski. 1978. "A Study of the Efficacy of Simulated Laboratory Experiments." *Journal of Chemical Education* 55: 602–604.

Cavin, C.S. and J.J. Lagowski. 1978. "Effects of Computer Simulated or Laboratory Experiments and Student Attitudes on Achievement and Time in a College General Chemistry Laboratory Course." *Journal of Research in Science Teaching* 15: 455–463.

Champagne, A.B., L.E. Klopfer, and R.F. Gunstone. 1982. "Cognitive Research and the Design of Science Instruction." *Educational Psychologist* 17: 31–53.

Chi, M.T.H., P.J. Feltovich, and R. Glaser. 1981. "Categorization and Representation of Physics Problems by Experts and Novices." *Cognitive Science* 5: 121–152.

Chi, M.T.H. and R. Glaser. 1981. "The Measurement of Expertise: Analysis of the Development of Knowledge and Skills as a Basis for Assessing Achievement." In E.L. Baker and E.S. Quellmalz, eds. *Design, Analysis and Policy in Testing*. Beverly Hills, Calif.: Sage Publications.

Clark, R.W. 1972. *Einstein: The Life and Times*. New York, N.Y.: Avon.

Clement, J.J. 1982. "Students' Preconceptions in Introductory Mechanics." *American Journal of Physics* 50: 66–71.

Clement, J. 1985. "Misconceptions in Graphing." In Proceedings of the Ninth International Conference of the Group on the Psychology of Mathematics Education. Noordwijkerhout, The Netherlands.

Clement, J. 1987. "Overcoming Students' Misconceptions in Physics: The Role of Anchoring Intuitions and Analogical Validity." In J.D. Novak, ed. *Proceedings of the Second International Seminar on Misconceptions and Educational Strategies in Science and Mathematics*, Vol III. Ithaca, N.Y.: Department of Education, Cornell University. pp. 84–97.

Cohen, H.D., D.F. Hillman, and R.M. Agne. 1978. "Cognitive Level and College Physics Achievement." *American Journal of Physics* 46: 1026–1029.

College Board–Educational Testing Service. 1989. ALGEBRIDGE Project. Princeton, N.J.: College Entrance Examination Board–Educational Testing Service. (To be published by Janson Publications.)

College Board, 1983. *Academic Preparation for College: What Students Need to Know and Be Able to Do*. New York: College Entrance Examination Board.

Collins, A., J.S. Brown, and S.E. Newman. 1989. "Cognitive Apprenticeship: Teaching the Crafts of Reading, Writing and Mathematics." In L.B. Resnick, ed., *Knowing, Learning and Instruction*. Hillsdale, N.J.: Lawrence Erlbaum Associates.

De Avila, E.A. 1988. "Bilingualism, Cognitive Function, and Language Minority Group Membership." In R. Cocking and J. Mestre, eds. *Linguistic and Cultural Influences on Learning Mathematics*. Hillsdale, N.J.: Lawrence Erlbaum Associates. pp. 101–122.

Dossey, J.A., I.V.S. Mullis, M.M. Lindquist, and D.L. Chambers. 1988. "The Mathematics Report Card: Are We Measuring Up?" *Trends and Achievement Based on the 1986 National Assessment*. Princeton, N.J.: Educational Testing Service.

Doyle, W. 1983. "Academic Work." *Review of Educational Research* 53: 159–199.

Driver, R. 1983. *The Pupil as Scientist?* London, United Kingdom: The Open University Press.

Dubravcic, M.F. 1979. "Practical Alternatives to Laboratory in a Basic Chemistry Course." *Journal of Chemical Education* 56: 235–237.

Eylon, B.S. and F. Reif. 1984. "Effect of Knowledge Organization on Task Performance." *Cognition and Instruction* 1: 5–44.

Fensham, P. and A. Kornhauser. 1982. "Challenges for the Future of Chemical Education." In M. Gardner, ed. Sixth International Conference on Chemical Education. College Park, Md.: University of Maryland. pp. 115–137.

Feyerabend, P. 1978. *Against Method: Outline of an Anarchistic Theory of Knowledge*. London, U.K.: Verso.

Fredette, N.H. and J.J. Clement. 1981. "Student Misconceptions of an Electric Circuit: What Do They Mean?" *Journal of College Science Teaching* March: 280–285.

Frederiksen, N. 1984. "The Real Test Bias: Influences of Testing on Teaching and Learning." *American Psychologist* 39: 193–202.

Fuller, R.G., R. Karplus, and A.E. Lawson. 1977. "Can Physics Develop Reasoning?" *Physics Today* 30: 1383–1388.

The Fundamental Particles and Interactions Chart Committee. 1988. "Fundamental Particles and Interactions." *The Physics Teacher* December: pp. 536–565.

Glaser, R. 1988. "Cognitive and Environmental Perspectives on Assessing Achievement." *Assessment in the Service of Learning*. Princeton, N.J.: Educational Testing Service.

Glen, W. 1982. *The Road to Jamarillo: Critical Years of the Revolution in Earth Science*. Stanford, Calif.: Stanford University Press.

Goldberg, F.M. and L.C. McDermott. 1986. "Student Difficulties in Understanding Image Formation by a Plane Mirror." *The Physics Teacher* November: 472–480.

Goldberg, F.M. and L.C. McDermott. 1987. "An Investigation of Student Understanding of the Real Image Formed by a Converging Lens or Concave Mirror." *American Journal of Physics* 55: 108–119.

Goodlad, John. 1984. *A Place Called School*. New York: McGraw-Hill.

Gowin, B.D. and J.D. Novak. 1984. *Learning How to Learn*. New York: Cambridge University Press.

Griffiths, G.H. 1976. "Physics Teaching: Does It Hinder Intellectual Development?" *American Journal of Physics* 44: 81–85.

Hallam, A. 1973. *A Revolution in the Earth Sciences: From Continental Drift to Plate Tectonics*. Oxford, U.K.: Clarendon Press.

Halloun, I.A. and D. Hestenes. 1987. "Modeling Instruction in Mechanics." *American Journal of Physics* 55: 455–462.

Hardiman, P.T., R. Dufresne, and J.P. Mestre. 1989. "The Relation between Problem Categorization and Problem Solving among Experts and Novices." *Memory and Cognition* 17: 627–638.

Harrison, E. 1989. *Darkness at Night: A Riddle of the Universe*. Cambridge, Mass.: Harvard University Press.

Hawking, S.W. 1988. *A Brief History of Time: From the Big Bang to Black Holes*. New York: Bantam Books.

Hayes, J.R. and L.S. Flower. 1986. "Writing Research and the Writer." *American Psychologist* 41: 1106–1113.

Helgeson, S.L., R.E. Stake, and I.E. Weiss. 1977. *The Status of Pre-*

College Science, Mathematics and Social Science Education: 1955–1975, Vol. I. Educational Resources Information Center (ED 166 034).

Heller, J.I. and F. Reif. 1984. "Prescribing Effective Human Problem Solving Processes: Problem Description in Physics." *Cognition and Instruction* 1:177–216.

Helm, H. and J.D. Novak, eds. 1983. *Proceedings of the International Seminar on Misconceptions in Science and Mathematics.* Ithaca, N.Y.: Department of Education, Cornell University.

International Association for the Evaluation of Educational Achievement. 1988. *Science Achievement in Seventeen Countries.* Oxford, U.K.: Pergamon Press.

Janvier, C, ed. 1987. *Problems of Representation in the Teaching and Learning of Mathematics.* Hillsdale, N.J.: Lawrence Erlbaum Associates.

Judson, H.F. 1979. *The Eighth Day of Creation: Makers of the Revolution in Biology.* New York: Simon & Schuster.

Kinnear, J. 1983. "Identification of Misconceptions in Genetics and the Use of Computer Simulations in their Correction." In H. Helm and J. Novak, eds., *Proceedings of the International Seminar on Misconceptions in Science and Mathematics.* Ithaca, N.Y.: Department of Education, Cornell University. pp. 84–92.

Kruglak, H. 1952. "Experimental Outcomes of Laboratory Instruction in Elementary College Physics." *American Journal of Physics* 20: 136–141.

Kruglak, H. 1953. "Achievement of Physics Students with and without Laboratory Work." *American Journal of Physics* 21: 14–16.

Kuhn, T. 1970. *The Structure of Scientific Revolutions.* Chicago, Ill.: University of Chicago Press.

Lightman, A.P., J.D. Miller, and B.J. Leadbeater. 1987. "Contemporary Cosmological Beliefs." In J.D. Novak, ed. *Proceedings of the Second International Seminar on Misconceptions and Educational Strategies in Science and Mathematics*, Vol. III. Ithaca, N.Y.: Department of Education, Cornell University. pp. 309–321.

Linn, M.C. 1986. "Science." In R. Dillon and R. Sternberg, eds. *Cognition and Instruction.* Orlando, Fla.: Academic Press. pp. 155–204.

Lipson, J.I. 1988. "Testing in the Service of Learning Science: Learning-Assessment Systems that Promote Educational Excellence and Equality." *Assessment in the Service of Learning.* Princeton, N.J.: Educational Testing Service.

Lochhead, J. and J.P. Mestre. 1988. "From Words to Algebra: Mending Misconceptions." In A.F. Coxford and A.P. Shulte, *The Ideas of Algebra, K-12:* 1988 Yearbook of the National Council of Teachers of

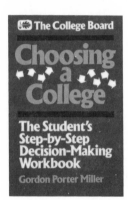

Colleges • Costs • Majors

The College Handbook, 1989-90

Recommended by college admissions officers more than any other guide, the *Handbook* contains detailed descriptions supplied by more than 3,100 two- and four-year colleges. It provides essential facts on fall 1990 admissions requirements; size, location, and campus setting; majors offered; annual expenses; financial aid; student activities; campus life; and more. Also: more than 30 indexes to colleges by special features and authoritative guidance on making informed college decisions.
003365 ISBN: 0-87447-336-5, 1989, 2,000 pages, $17.95

Index of Majors, 1989-90

An easy-to-use reference for those who know what they want to study but not where. It lists 500 majors and the more than 3,000 colleges, universities, and graduate schools, state by state, that offer them. Degree levels—certificate, associate, bachelor's, master's, doctor's, and first professional—at which these majors are offered are provided. Also: information on special academic programs such as independent study, student-designed majors, cooperative education, and combined bachelor's/graduate degrees.
003373 ISBN: 0-87447-337-3, 1989, 790 pages, $14.95

The College Cost Book, 1989-90

A step-by-step guide to help families plan ahead for college costs, apply for financial aid, and make the most of personal resources. It contains the College Board's official advice concerning financial aid and detailed facts about costs and the availability of both need-based aid and scholarships at more than 3,100 colleges. Also: worksheets, cost tables, and lists of colleges that offer special payment plans and tuition/fee waivers.
003381 ISBN: 0-87447-338-1, 1989, 280 pages, $13.95.

SPECIAL 36% DISCOUNT
Order *The College Handbook, Index of Majors,* and *The College Cost Book as a set* for just $29.95. It's like getting one book FREE!
239335, $29.95

Order Form
for College Board Publications

Mail to: Department N74
College Board Publications
Box 886
New York, New York 10101-0886

Orders for five or more copies of a single item receive a 20% discount.* The College Board pays fourth class book rate postage on prepaid orders. Postage will be charged on orders received on purchase orders or requesting a faster method of shipment. Allow 30 days from receipt of order for delivery. (For faster delivery when prepaying, call 717-348-9288 for the cost of UPS postage to include in your check.)

Enclosed is ☐ my check, made payable to the College Board, in the noted amount, or
☐ an institutional purchase order.

Qty.	*Item No.*	*Title*	*Price*	*Amount*
______	003535	Student's Guide to Good Writing	$9.95	$______
______	003330	Choosing a College	9.95	______
______	003578	College Board Guide to High Schools†	89.95	______
______	003357	Countdown to College	9.95	______
______	003179	Campus Health Guide	14.95	______
______	003365	The College Handbook, 1989-90	17.95	______
______	003373	Index of Majors, 1989-90	14.95	______
______	003381	The College Cost Book, 1989-90	13.95	______
		SPECIAL SAVINGS		
______	239335	1989-90 Three-Book Set: The College Handbook, Index of Majors, The College Cost Book	29.95	______
______	295740	Admissions Reference Bookshelf	55.00	______
______	002474	Your College Application	9.95	______
______	002571	Writing Your College Application Essay	9.95	______
______	002601	Campus Visits and College Interviews	9.95	______
______	003225	Summer on Campus	9.95	______
______	003047	College Bound	9.95	______
______	003160	The College Guide for Parents	12.95	______
______	002482	How to Pay for Your Children's College Education	12.95	______
______	002490	College to Career	9.95	______

*(Less discount, if any) Subtotal $______
Calif. residents add 6.25 sales tax (thru 12/31/90; after, 6%) $______
Pa. residents add 6% sales tax $______
Handling charge $______
Grand Total $______

†to ship April 1990

Please print:

Name ______________________ Title ______________________

Institution (if any) ______________________

Street Address ______________________

City ______________ State ______________ Zip Code ______________

218128

Mathematics. Reston, Va.: National Council of Teachers of Mathematics. pp. 127–135.

Malone, D., A. Naiman, and R. Tinker. 1984. *Experiments in Chemistry: Teaching Guide*. Pleasantville, N.Y.: H.R.M. Software.

Mandinach, E.B. and M. Thorpe. 1987. *The Systems Thinking and Curriculum Innovation Project*. Technical Report 87-6, Part I. Cambridge, Mass.: Educational Technology Center, Graduate School of Education, Harvard University.

McCloskey, M., A. Caramazza, and B. Green. 1980. "Curvilinear Motion in the Absence of External Forces: Naive Beliefs about the Motion of Objects." *Science* 210: 1139–1141.

McCloskey, M. 1983. "Intuitive Physics." *Scientific American* 248: 122.

McDermott, L.C., L.K. Piternick, and M.L. Rosenquist. 1980. "Helping Minority Students Succeed in Science. Part I: Development of a Curriculum in Physics and Biology." *Journal of College Science Teaching* 9: 135–140.

McDermott, L.C., L.K. Piternick, and M.L. Rosenquist. 1980. "Helping Minority Students Succeed in Science. Part II: Implementation of a Curriculum in Physics and Biology." *Journal of College Science Teaching* 9: 201–205.

McDermott, L.C., L.K. Piternick, and M.L. Rosenquist. 1980. "Helping Minority Students Succeed in Science. Part III: Requirements for the Operation of an Academic Program in Physics and Biology." *Journal of College Science Teaching* 9: 261–265.

McDermott, L.C. 1984. "Research on Conceptual Understanding in Mechanics." *Physics Today* 37(7): 24–32.

Mestre, J.P. 1987. "Why Should Mathematics and Science Teachers Be Interested in Cognitive Research Findings?" *Academic Connections* Summer Issue: 3–5, 8–11.

Mestre, J. 1988. "The Role of Language Comprehension in Mathematics and Problem Solving." In R. Cocking and J. Mestre, eds. *Linguistic and Cultural Influences on Learning Mathematics*. Hillsdale, N.J.: Lawrence Erlbaum Associates. pp. 201–220.

Mestre, J. and J. Touger. 1989. "Cognitive Research: What's in it for Physics Teachers." *The Physics Teacher* 27 (Sept.): 447–456.

Mestre, J.P., R. Dufresne, W. Gerace, P. Hardiman, and J. Touger. 1989. "Enhancing Higher-Order Thinking Skills in Physics" Internal Report 178. Scientific Reasoning Research Institute, University of Massachusetts at Amherst.

Minstrell, J. 1982. "Explaining the 'At Rest' Condition of an Object." *The Physics Teacher* 20: 10.

Minstrell, J. 1987. *Classroom Dialogues for Promoting Physics Under-*

standing. Paper Presented at the Annual Meeting of the American Educational Research Association, Washington, D.C.

Murnane, R.J. and S.A. Raizen. 1988. *Improving Indicators of the Quality of Science and Mathematics Education in Grades K-12*. Washington, D.C.: National Academy Press.

National Commission on Excellence in Education. 1983. *A Nation at Risk*. Washington, D.C.: U.S. Government Printing Office.

National Research Council. 1989. *Everybody Counts: A Report to the Nation on the Future of Mathematics Education*. Washington, D.C.: National Academy Press.

National Science Board Task Committee on Undergraduate Science and Engineering Education. 1986. *Undergraduate Science, Mathematics and Engineering Education*. (NSB 86–100). Washington, D.C.: NSB.

Nickerson, R.S. 1982. "Notes about Reasoning." Report #5191. Bolt, Beranek & Newman, Cambridge, Mass.

Nickerson, R.S., D.N. Perkins, and E.E. Smith. 1985. *The Teaching of Thinking*. Hillsdale, N.J.: Lawrence Erlbaum Associates.

Novak, J.D., ed. 1987. *Proceedings of the Second International Seminar on Misconceptions and Educational Strategies in Science and Mathematics*, Vol. III. Ithaca, N.Y.: Department of Education, Cornell University.

Palincsar, A.S. and A.L. Brown. 1984. "Reciprocal Teaching of Comprehension-Fostering and Monitoring Activities." *Cognition and Instruction* 1: 117–175.

Papert, S. 1980. *Mindstorms*. New York: Basic Books.

Phillips, M. 1981. "Early History of Physics Laboratories for Students at the College Level." *American Journal of Physics* 49: 522–527.

Putnam, R.T. 1987. "Structuring and Adjusting Content for Students: A Study of Live and Simulated Tutoring of Addition." *American Educational Research Journal* 24: 13–48.

Raphael, T.E. 1987. "Research on Reading: 'But What Can I Teach on Monday?'" In V. Richardson-Koehler, ed. *Education Handbook: A Research Perspective*. New York: Longman.

Resnick, L.B. 1983. "Mathematics and Science Learning: A New Conception." *Science* 220: 477–478.

Resnick, L.B. 1987. *Education and Learning to Think*. Washington, D.C.: National Academy Press.

Robinson, M.C. 1979. "Undergraduate Laboratories in Physics: Two Philosophies." *American Journal of Physics* 47: 859–862.

Rock, Donald A., et al. 1984. "Factors Associated with Test Score Decline: Briefing Paper." Princeton, N.J.: Educational Testing Service.

Rosen, S. 1954. "A History of the Physics Laboratory in the American

Public High School (to 1910)." *American Journal of Physics* 22: 194–204.
Rowe, M.B. 1973. *Teaching Science as Continuous Inquiry.* New York: McGraw-Hill.
Sadler, P.M. 1987. "Misconceptions in Astronomy." In J.D. Novak, ed. *Proceedings of the Second International Seminar on Misconceptions and Educational Strategies in Science and Mathematics*, Vol III. Ithaca, N.Y.: Department of Education, Cornell University. pp. 422–425.
Seyfert, C.K. and L.A. Sirkin. 1979. *Earth History and Plate Tectonics: An Introduction to Historical Geology*, 2nd ed. New York: Harper & Row.
Shamos, M.H. 1987. *Great Experiments in Physics: Firsthand Accounts from Galileo to Einstein.* New York: Dover.
Shymansky, J.A., W.C. Kyle, and J. Alport. 1982. "Research Synthesis on the Science Curriculum Projects of the Sixties." *Educational Leadership* October: 63–66.
Strauss, N.J. and T. Fulwiler. 1987. "Interactive Writing and Learning Chemistry." *Journal of College Science Teaching* 16: 256–262.
Sweller, J. 1988. "Cognitive Load During Problem Solving: Effects on Learning." *Cognitive Science* 12: 257–285.
Sweller, J. and G.A. Cooper. 1985. "The Use of Worked Examples as a Substitute for Problem Solving in Learning Algebra." *Cognition and Instruction* 2: 59–89.
Szamosi, G. 1986. *The Twin Dimensions: Inventing Space and Time.* New York, N.Y.: McGraw-Hill.
Task Force on Education for Economic Growth. 1984. *Action in the States: Progress toward Education Renewal*, A Report by the Task Force on Education for Economic Growth. Denver, Colo.: Education Commission of the States.
Tinnesand, M. and A. Chan. 1987. "Step 1: Throw Out the Instructions." *The Science Teacher* September: 43–45.
Trefil, J.S. 1983. *The Moment of Creation: Big Bang Physics from Before the First Millisecond to the Present Universe.* New York: Scribner.
U.S. Department of Education, National Center for Education Statistics. 1982. *Digest of Education Statistics: 1982.* Washington, D.C.: U.S. Government Printing Office.
Uyeda, S. 1978. *The New View of the Earth.* San Francisco, Calif.: Freeman.
Wandersee, J.H. 1983. "Students' Misconceptions about Photosynthesis: A Cross-Age Study." In H. Helm and J. Novak, eds. *Proceedings of the International Seminar on Misconceptions in Science and*

Mathematics. Ithaca, N.Y.: Department of Education, Cornell University. pp. 441–465.

Watson, J.D. 1968. *The Double Helix*. New York: Atheneum.

Wetzel, J.R. 1987, June. *American Youth: A Statistical Snapshot*. Report submitted to the National Science Board by the William T. Grant Foundation Commission on Work, Family and Citizenship.

Whimbey, A. and J. Lochhead. 1984. *Beyond Problem Solving and Comprehension*. Hillsdale, N.J: Lawrence Erlbaum Associates.

Whimbey, A. and J. Lochhead. 1987. *Problem Solving and Comprehension*, 4th ed. Hillsdale, N.J.: Lawrence Erlbaum Associates.

White, B. and P. Horwitz. 1987. "Thinkertools: Enabling Children to Understand Physical Laws." Technical Report #6470. Bolt, Beranek and Newman, Cambridge, Mass.

Wolf, D.P. 1988. *Reading Reconsidered*. New York, N.Y.: College Entrance Examination Board.

Appendix

A Biology Example of Structured Inquiry

This appendix is provided to give teachers additional illustration of the structured inquiry approach, in this case as applied in biology. We will use enzymatic reactions as our example.

A convenient and inexpensive enzymatic reaction for students to investigate involves the enzyme phenol oxidase and the substrate catechol. (Pyrocatechol can be substituted for the substrate.) In the reaction we will study, catechol is converted to O-benzoquinone. Although this reaction does not occur in living cells, it is similar to one that does. Catechol is a chemical analog of the chemical substrate used in living systems, tyrosine. In living systems, phenol oxidase converts tyrosine to the brown/black pigment melanin. Melanin is the pigment that gives a dark color to the skin and hair of animals. The details of the reaction are unimportant. In both cases, the final product has a dark brown/black color.

The enzyme phenol oxidase can be extracted from a raw potato by grinding it in water. This is most easily done in a blender, but could be done by finely chopping the potato and adding water. The material should be strained through cheesecloth or some other suitable sieve. The substrate, catechol, is a colorless liquid, but the final product is black. Therefore, the progress of the reaction can be monitored by a color change from a light red-orange, to reddish brown, to brown, to brown-black, to black. The darker the color of the liquid, the greater the amount of product formed.

Approaching Scientific Questions Experimentally

- *Outcome A: Sufficient familiarity with laboratory and fieldwork to ask appropriate scientific questions and to recognize what is involved in experimental approaches to the solution of such questions.*

A.1 Identifying and posing meaningful, answerable scientific questions.
A.2 Formulating working hypotheses.
A.3 Selecting suitable methods for answering scientific questions or testing hypotheses.
A.4 Designing appropriate procedures for laboratory or fieldwork to test hypotheses.

The teacher could begin by indicating that the focal topic is the nature of enzymatic reactions and start by engaging the class in a discussion of the nature of reactions in general, comparing and contrasting the nature of enzymatic and nonenzymatic reactions. The students are likely to have more experience with nonenzymatic reactions, which proceed faster at increasing temperatures. In contrast, enzymatic reactions proceed fastest at some temperature that is optimal for that particular enzyme and will stop altogether at certain temperature extremes. The discussion should lead students to consider the consequences of the general method by which enzymes work: That is, an enzyme facilitates a reaction by binding the substrates, then releases the product(s), and is recycled.

During the discussion, students should identify and pose meaningful, answerable scientific questions, such as: Will an increase in temperature always increase the reaction rate? Will a change in pH alter the reaction rate? Is an enzyme used up in the reaction? Is an enzyme converted to a product? This part of the discussion should be used to help students recognize the difference between a general scientific question and a specific hypothesis. A *hypothesis* is a general statement that includes a prediction. Students should be guided to use their questions to form working hypotheses. For example, the questions "are enzymes recycled?" and "how does temperature affect reaction rates?" can be restated as the following two hypotheses (only the first of which is valid): 1) The enzyme is recycled rather than being converted to product; thus, varying the enzyme concentration should not influence the total amount of product formed but would influence the length of time it takes to convert all the substrate to product. 2) The rate of all chemical reactions will increase with increasing temperature.

Some guidance may be needed as students design experiments that will answer their scientific questions or test their hypotheses.

Students whose questions or hypotheses dealt with the effects of temperature or pH on the rate of chemical reactions will probably have no difficulty designing experiments to determine the effects of these factors. Designing a method for determining whether the enzyme is used up or converted to product is more challenging. However, with thought, students should be able to design an experimental test of this by diluting the enzyme. A suitable hypothesis might predict that regardless of enzyme concentration, the quantity of final product will be the same, although the length of time for the conversion of all the substrate to product would be longer with less enzyme.

Students should be encouraged to plan their experiments before beginning work. For example, they need to consider:

1. How the process of the reaction will be followed or measured.
2. How frequently measurements will be made.
3. The time period for measurement.
4. Suitable controls.
5. The temperature range to be explored.
6. How changes in temperature will be effected.
7. How pH will be altered.
8. How much enzyme should be diluted.
9. How many different concentrations of the enzyme should be tested.

Gathering Scientific Information

- *Outcome B: The skills to gather scientific information through laboratory, field, and library work.*

B.1 Observing objects and phenomena.
B.2 Describing observations accurately using appropriate language.
B.3 Assembling the appropriate measuring instruments.
B.4 Measuring objects and changes quantitatively.
B.5 Analyzing observational and experimental data.
B.6 Developing sound skills in using common laboratory and field equipment.
B.7 Performing common laboratory techniques with care and safety.

To develop the skills that fall under outcome B, students should observe the progress of the reaction in a sample containing enzyme and substrate maintained at room temperature. If this reaction is observed carefully, it will be noted that the color changes more quickly at the surface and near any bits of potato that are floating in the test tube. The teacher should guide the students in realizing that, because of this, readings of color change will be more accurate if the test tube is swished before the color change is assessed.

Under outcome B.2, students should be encouraged to decide on a method of quantifying the color change. One possible scheme might be:

− = no color change
\+ = very slight color change (reddish brown)
\+ + = light color change (brown)
\+ + + = definite color change (brown-black)
\+ + + + = deep color (black)

Outcome B also requires that the students recognize the need for controls and that they decide on the appropriate comparison for each of their experiments. One control might be a sample containing substrate, but lacking the enzyme. Other controls would depend on the particular experiment. If the effect of pH is being investigated, a control might be a sample run at a neutral pH. If the effect of temperature is under investigation, a control might be a sample run at room temperature.

Throughout both the discussion and experimental phases, the teacher should reinforce key procedures as the need arises, such as organizing the experiments with a table or flow chart, developing methods for safe and accurate measurements of pH and temperature, and careful labeling of samples and data. However, some procedures, such as the proper use of safety glasses, should not be left for students to discover on their own.

Organizing and Communicating Results

▪ *Outcome C: The ability to organize and communicate the results obtained by observation and experimentation.*

C.1 Organizing data and observations.

C.2 Presenting data in the form of functional relationships.
C.3 Extrapolating functional relationships beyond actual observations, when warranted, and interpolating between observations.

When organizing data (outcome C.1), students should carefully consider which samples are to be compared. If their original hypothesis was that pH would influence reaction rate, comparisons should be made between the control and the samples in which pH was varied; if the original hypothesis was that temperature would increase reaction rate, comparisons should be made between the control and the samples in which temperature was varied. If the hypotheses involved the enzyme, then comparisons should be among samples with different enzyme concentrations.

An important point to convey to students is that experiments should be planned and organized so that only one variable is considered at a time. If time permits, this can be left to students to discover after they have completed the first round of experiments. Those who did not control variables may have to repeat some or all of their measurements.

Outcome C.2 might involve graphing the results (observations) that were recorded in tabular form during the experiment, e.g., plotting the degree of color change versus time.

Outcome C.3, extrapolating, can lead to a particularly lively class discussion. Phenomena such as the temperature dependence of enzymatic reactions are good illustrations of the limits and dangers of this procedure.

Drawing Conclusions

- *Outcome D: The ability to draw conclusions or make inferences from data, observation, and experimentation, and to apply mathematical relationships to scientific problems.*

D.1 The ability to interpret data presented in tables and graphs.
D.2 The ability to interpret, in nonmathematical language, the relationships presented in mathematical form.
D.3 Evaluating a hypothesis in view of observations and experimental data.

D.4 Formulating appropriate generalizations, laws, or principles warranted by the relationships found.

Students should be encouraged to discuss and critique various groups' efforts to interpret the data they collected and graphed. One prediction might have been that the amount of product formed would continuously increase over time. When students examine their data in either tabular or graphic form, they should realize that the product initially increases over time as substrate is enzymatically converted to product. However, after the reaction has continued for a period of time, the curve depicting the amount of product formed versus time levels off. Students may offer several explanations, each of which should be fully considered before allowing anyone to conclude that all or most of the substrate has been converted to product. This understanding may prompt some students to develop additional hypotheses to test. For example, "If the curve levels off when all the substrate has been converted to product, then adding more substrate should result in more product being formed." A new experiment could easily be designed to test this hypothesis. Students should also be guided to recognize that when the amount of product is plotted against time, the slope of the curve before it levels off indicates how quickly substrate is being converted to product (i.e., the rate of reaction).

Students should be encouraged to evaluate their hypotheses in light of the data (outcomes D.1 and D.2). If the students' initial hypothesis was that the rate of reaction would increase with temperature, they should compare reaction rates among the samples in which temperature was varied. As a result of this comparison, students will notice that their hypothesis is only partially true (outcome D.3). They will need to develop a new hypothesis that is consistent with all of the experimental results, for example, that the reaction rate increases unless the temperature exceeds a critical value.

Students might then be encouraged to develop hypotheses to explain why extreme temperatures, such as those caused by boiling, would either dramatically reduce the product formed or prevent the reaction. These may lead to new experiments to test the hypotheses. For example, hypotheses might include: 1) no product forms because boiling destroys the substrate or 2) no product forms because boiling destroys the enzyme. (Only the second hypothesis should be supported by results of subsequent experiments.)

The students' initial scientific questions may have concerned the fate of the enzyme during the reaction. They may have asked whether the enzyme is: converted to product, destroyed in converting the substrate to product, or recycled in an unchanged form. The experiments might then have varied the concentration of enzyme to determine whether this influenced either the reaction rate or the amount of product formed. In this case, students would compare the reaction rates (slopes of the initial part of the curves of product versus time) of samples with different enzyme concentrations, as well as the height of the curves when they level off (amount of product formed) (outcome D.1). Students should be able to view their results as an indication that enzyme concentration alters the reaction rate, but that it does not alter the total amount of product formed (outcomes D.1 and D.2). With this information, students should be able to decide whether the data are consistent with their initial hypotheses (D.3). The data might be generalized by the statement: "Enzyme concentration influences the reaction rate, but not the final amount of product formed." The teacher should encourage the students to consider the implications of this observation. If the amount of enzyme does not influence the amount of product formed, can it be used up during the reaction? If enzyme concentration does influence how quickly substrate is converted to product, is this consistent with the hypothesis that the enzyme is recycled?

Recognizing the Role of Experimental Work in Constructing Theories

- *Outcome E: The ability to recognize the role of observation and experimentation in the development of scientific theories.*

E.1 Recognizing the need for a theory to relate different phenomena and empirical laws or principles.

E.2 Formulating a theory to accommodate known phenomena and principles.

E.3 Specifying phenomena and principles that are satisfied or explained by a theory.
E.4 Deducing new hypotheses from a theory and directing experimental work to test it.
E.5 Formulating, when warranted by new experimental findings, a new revised, refined, or extended theory.

After exploring various factors influencing the rate of an enzymatic reaction, it should be easy to convince students of the value of a general "theory or set of postulates" to relate the diverse facets of their investigations (outcome E.1). Formulating a theory from this limited set of experiments (E.2) will be more difficult. Students may be led to generate postulates such as:

1. Enzymes are catalysts, substances that alter the rate of a reaction without being consumed in the process.
2. Enzymes do not alter how much product can be formed, only the rate at which product is formed.
3. Extreme conditions, such as boiling, inactivate enzymes.

It is at this point that a model for enzyme activity might be introduced. The "lock-and-key" model is common. In this model, some region on the surface of the enzyme is pictured as matching the shape of the substrate, much as a lock matches its key. Students can relate their observations of the slowing or cessation of the reaction that accompanies changes in pH or temperature to changes in the shape of the enzyme that would be caused by these conditions. Likewise, the effect of enzyme concentration on the reaction is consistent with the model's portrayal of enzyme molecules as combining with the substrate, facilitating the reaction and then being released for reuse.

These postulates and models would, of course, generate new hypotheses that could be tested by additional experiments (outcome E.4). For example, the lock-and-key model leads to the hypothesis that enzymes should be specific for certain substrates. Students could devise experiments to examine whether this enzyme, (phenol oxidase) increases the rate of other reactions. In addition, they may hypothesize that the "theory" applies to all enzymatic reactions and begin to test other enzymatic reactions to determine whether they share the same important properties. The postulates may be revised to accommodate new data (outcome E.5). For example, if the students studied an enzyme found in organisms commonly living in hot springs, they might find that enzymatic activity continues at high temperatures but that low temperatures inactivate the enzyme.

This series of experiments might even be related to broader questions centered on the biological significance of enzymes. Students might consider why enzymes govern the chemical reactions of cells. Perhaps they could explore whether the sensitivity of enzymatic reactions to factors such as pH and temperature is a reason for cells to regulate their internal environment carefully.

The teacher should note that classroom discussions will have to provide the student with some understanding of how enzymes work before the student can be expected to form hypotheses about enzyme activity and the factors that may influence it. For example, as a result of these discussions, students should realize that most of the chemical reactions within cells involve enzymes and that enzymes make these reactions possible under conditions within the cell. The students should know that enzymes work by binding to the reacting molecules(s), (*substrate*) thereby increasing the likelihood of producing the product(s). This binding depends on the shape of the enzyme and the substrate(s) because the substrate(s) must fit into a specific region of the enzyme molecule. The student should also know already that after the reaction occurs, the product(s) is released from the enzyme, which is then freed to catalyze the reaction of other molecules. In other words, the enzyme is recycled. The student should know that enzymatic reactions proceed most quickly at some optimal temperature and pH, and that deviating from these optimal values usually slows the rate of reaction. Reaction rate is slowed under temperature or pH conditions that alter the shape of the enzyme, reducing the closeness of fit between the enzyme and substrate. The reason the enzyme's shape is altered by nonoptimal pH and temperature is that enzymes are proteins and both pH and temperature influence the bonds that hold the protein in its characteristic shape. Extreme heat will destroy the enzyme and stop the reaction. However, students do not have to know why pH and temperature influence reaction rate to explore the effects of these factors on the rate of reaction.

Members of the Council on Academic Affairs, 1988–89

Mildred Alpern, History and Social Science Teacher, Spring Valley Senior High School, Spring Valley, New York (*Chair*, 1987–89)
Jeremy Kilpatrick, Professor of Mathematics Education, University of Georgia, Athens (*Vice Chair*, 1987–89)
Thomas R. Beyer, Jr., Chair of Foreign Language Division, Middlebury College, Middlebury, Vermont (1987–90)
Iris Gonzalez, College Board Student Representative, Roy Miller High School, Corpus Christi, Texas (1987–90)
John Howe, Professor of History, University of Minnesota, Minneapolis (1988–91)
Derrick L. Jones, College Board Student Representative, Tufts University, Medford, Massachusetts (1988–91)
Stanley N. Katz, President, American Council of Learned Societies, New York (1987–90)
William Larkin, Assistant Superintendent, Division of Curriculum and Instruction, Milwaukee, Wisconsin (1987–90)
Marie Lerner-Sexton, Choral Music Teacher, Shawnee Mission South High School, Shawnee Mission, Kansas (1987–90)
Naomi Martin, Coordinator of Secondary Mathematics/Science, Horry County School District, South Carolina (1986–89)
Dorsey T. Patterson, Principal, Booker T. Washington High School, Memphis, Tennessee (1987–90)
Claire Pelton, Director of Curriculum/Testing, San Jose Unified School District, San Jose, California (1988–90)
Linda B. Salamon, Dean, College of Arts and Sciences, Washington University, St. Louis, Missouri (1988–89)
Judith Sanford-Harris, Assistant Dean of Academic Affairs, Bunker Hill Community College, Boston, Massachusetts (1986–89)
Jane C. Schaffer, English Department Chair, West Hills High School, Santee, California (1988–91)
Phillip Uri Treisman, Director, Undergraduate and Professional Development Programs, University of California, Berkely, California (1987–90)